T0226090

Springer Series on Naval Architecture, Marine Engineering, Shipbuilding and Shipping

Volume 11

Series Editor

Nikolas I. Xiros, University of New Orleans, New Orleans, LA, USA

The Naval Architecture, Marine Engineering, Shipbuilding and Shipping (NAMESS) series publishes state-of-art research and applications in the fields of design, construction, maintenance and operation of marine vessels and structures. The series publishes monographs, edited books, as well as selected PhD theses and conference proceedings focusing on all theoretical and technical aspects of naval architecture (including naval hydrodynamics, ship design, shipbuilding, shipyards, traditional and non-motorized vessels), marine engineering (including ship propulsion, electric power shipboard, ancillary machinery, marine engines and gas turbines, control systems, unmanned surface and underwater marine vehicles) and shipping (including transport logistics, route-planning as well as legislative and economical aspects).

The books of the series are submitted for indexing to Web of Science.

All books published in the series are submitted for consideration in Web of Science.

More information about this series at http://www.springer.com/series/10523

Sidun Fang · Hongdong Wang

Optimization-Based Energy Management for Multi-energy Maritime Grids

 Springer

Sidun Fang 🆔
Chinese University of Hong Kong
Hong Kong
China

Hongdong Wang
School of Naval Architecture
Ocean and Civil Engineering
Shanghai Jiao Tong University
Shanghai, China

ISSN 2194-8445   ISSN 2194-8453 (electronic)
Springer Series on Naval Architecture, Marine Engineering, Shipbuilding and Shipping
ISBN 978-981-33-6736-4   ISBN 978-981-33-6734-0 (eBook)
https://doi.org/10.1007/978-981-33-6734-0

This Springer imprint is published by the registered company Springer Nature Singapore Pte Ltd.
The registered company address is: 152 Beach Road, #21-01/04 Gateway East, Singapore 189721, Singapore

For our beloved family and friends

Preface

Nowadays, the increment of international maritime trade decelerates by the influence of the global downside economy, and the even stricter environmental policies further intensify the competition between different sectors in maritime transportation systems, which motivates the electrification of all the attached sub-systems for higher energy efficiency, such as the all-electric ships, electrified ports, and various electrified ocean platforms. Different equipment and technologies have been integrated into those sub-systems, such as fuel cell, energy storage, gas capture system, alternative fuel, multi-energy management, and cold-ironing facilities, which gives birth to the maritime grids.

A typical maritime grid consists of generation, storage, and critical loads, and can operate either in grid-connected or in islanded modes, and operate under both the constraints of the energy system and the maritime transportation system, and formulates as *multi-energy maritime grids*. The energy management of this special system will shape the energy efficiency of the future maritime transportation system, and this is the main focus of this book.

This book mainly focuses on the energy management of the *multi-energy maritime grids*. With various practical cases, this book provides a cross-disciplinary view on green and sustainable shipping via the electrification of maritime grids.

Chapter 1 illustrates the background and motivation of the *multi-energy maritime grids*, and then the electrification trend of maritime grids is described; after that, different types of new technologies which are about to integrate are depicted and the concepts of *multi-energy maritime grids* are proposed after a comprehensive literature survey.

Chapter 2 briefly introduces the mathematical basics of optimization techniques used in this book, including the general optimization model, stochastic optimization model, robust optimization model, interval optimization model, convex optimization, and two optimization frameworks, i.e., two-stage optimization and bi-level optimization.

Chapter 3 illustrates the main management targets of *multi-energy maritime grids* under the background of extensive electrification. The targets include navigation tasks, energy consumption, gas emission, reliability under failures, lifecycle cost, and quality of service.

Chapter 4 introduces the formulation and solution of maritime grid optimization. At first, the classification is illustrated as (1) Synthesis optimization, (2) Design optimization, and (3) Operation optimization. Then different practical cases are given. At last, a compact formulation is proposed and a solution method is described.

Chapters 5–8 propose energy management models for maritime grids under different scenarios, i.e., under operating uncertainties (Chap. 5), energy storage integration (Chap. 6), multi-energy management (Chap. 7), and multi-source management (Chap. 8).

Chapter 9 concludes this book and gives the ways ahead. The key challenges are summarized and three promising problems are given: (1) data-driven technologies, (2) siting and sizing problems, and (3) energy management problems.

The intended audience of this book include

- Faculty, students, and researchers active in maritime transportation and interested in the environmental dimension of shipping.
- Carriers, shippers, infrastructure managers, and other logistics providers who aim at improving their environmental performance while staying in business.
- Technology designers and providers.
- Policy-makers at the national and international levels.
- Other stakeholders, environmental or other.

Hong Kong, China   Sidun Fang
October 2020

Acknowledgments

Earnest thanks to my Ph.D. supervisor Prof. Haozhong Cheng (Shanghai Jiao Tong University, Shanghai, China) for his supervision for 5 years. Thanks for the contribution of Dr. Qimin Xu (Shanghai Jiao Tong University, Shanghai, China) to the initial discussion, reviewing parts of the book. Special thanks also to Dr. Yue Song (The University of Hong Kong, Hong Kong, China), for contributing to the initial discussions, defining the scope of the book, and providing comments on parts of the book.

The authors also would like to thank a number of professionals who generously gave their time and provided comments on the draft chapters and draft sections of the book including Dr. Tianyang Zhao (Nanyang Technological University, Singapore), Dr. Shuli Wen (Shanghai Jiao Tong University, Shanghai, China), Dr. Shenxi Zhang (Shanghai Jiao Tong University, Shanghai, China), Dr. Ce Shang (Shanghai Jiao Tong University, Shanghai, China), Dr. Bin Gou (Southwest Jiao Tong University, Chengdu, China), and Dr. Yu Wang (Nanyang Technological University, Singapore).

Sincere gratitude also goes to the professionals, organizations, and institutions for permitting the authors to print some graphical material in the book, including Dr. Jean-Paul Rodrigue (Hofstra University, New York, USA), Dr. Anahita Molavia (Houston University, Houston, USA), and Prof. Josep M. Guerrero (Aalborg University, Aalborg, Denmark), and International Maritime Organization (IMO) and the Port of Los Angeles for providing valuable reports.

The authors are grateful for the funding support from the Key Laboratory of Marine Intelligent Equipment and System, Ministry of Education, Shanghai Jiao Tong University, and State Key Laboratory of Alternate Electrical Power System with Renewable Energy Sources, North China Electric Power University.

Contents

About the Authors

Dr. Sidun Fang born in 1991, received his B.E degree in School of Electrical Engineering, Chongqing University, Chongqing, China, in 2012, and his Ph.D. degree in Power System and its automation in the School of Electronic Information and Electrical Engineering, Shanghai Jiao Tong University, Shanghai, China, in 2017.

He serves as a Research Fellow in School of Electrical and Electronic Engineering, Nanyang Technological University, Singapore, till 2020. Now he is a postdoctoral fellow in The Chinese University of Hong Kong.

His research interests include optimal operation of mobile microgrids. Dr. Fang was awarded the Outstanding Graduate prize of Shanghai Jiao Tong University and his doctoral dissertation was nominated as Excellent Dissertation Papers in Shanghai Jiao Tong University in 2017. He has organized many special issues in journals and was invited as tutorial presenter in various top-tier international conferences. He also serves as the associate editor of International Transactions on Electrical Energy Systems.

Dr. Hongdong Wang born in 1989, received the degrees of undergraduate, master, and Ph.D. in Naval Architecture and Marine Engineering in Shanghai Jiao Tong University. Dr. Wang is now the associate professor and doctoral supervisor in the School of Naval Architecture, Ocean & Civil Engineering. Meanwhile, Dr. Wang is now the member of Youth Working Committee in the Chinese Society of Naval Architects and Marine Engineers and the director of China Association of the National Shipbuilding Industry in Shanghai. His research area focuses on the development and effectiveness evaluation of marine intelligent equipment, and the intelligent control based on the dynamic properties of marine equipment. More than 40 research papers in the abovementioned areas have been published, while 15 patents have been authorized. In addition, numerous projects are directed by Dr. Wang from National Natural Science Foundation of China, sub-program of National Key Research and Development Program of China, and so on. What's more, Dr. Wang is sponsored by China Association of Science and Technology Youth Talent Support Project and Shanghai Sailing Program.

Abbreviations

ABS	American bureau of shipping
AC/DC	Alternating current/Direct current
AES	All-electric ships
ANN	Artificial neural network
ARIMA	Autoregressive integrated moving average model
BOS	Battery only storage system
CCS	Carbon capture system
CHP/CCHP	Combined heat power/Combined cooling heat power
DG	Diesel engine or generator
DoD	Depth of discharge
dwt	Deadweight tonnes
ECA	Emission control area
EEDI	Energy efficiency design index
EEOI	Energy efficiency operating index
ELC	Equivalent life cycle
EMS	Energy management system
ESS	Energy storage system
EV	Electric vehicle
FC	Fuel cell
GHG	Greenhouse gas
HES	Heterogeneous energy storage
IMO	International maritime organization
LNG	Liquefied natural gas
LP	Linear optimization/Linear programming
MARPOL	International convention for the prevention of pollution from ships
MES	Multi-energy system
MILP	Mixed integer LP
MINLP	Mixed integer NLP
MIQP	Mixed integer QP
MSOC	Mean state of charge
NLP	Non-linear optimization/Non-linear programming
PTC	Power to thermal conversion

PV	Photovoltaic
QC	Quay crane
QCQP	Quadratic constrained QP
QoS	Quality of service
QP	Quadratic optimization/Quadratic programming
rpm	Round per minute
SDO	Synthesis-design-operation
SDP	Semi-definite programming
SOCP	Second-order cone programming
SVM	Supporting vector machine
UNCTAD	United nations conference on trade and development
UPS	Uninterruptable power supply

Chapter 1
Introduction to the Multi-energy Maritime Grids

1.1 Background and Motivation

1.1.1 Economy Growth and the Demand for Maritime Transport

After achieving a 3.1% growth in 2017, the growth rate of global economy declines to 3.0% in 2018 and further declines to 2.3% in 2019 [1]. In 2020 and afterward, a range of downside risks may further intensify the economy growth, such as the tariff between US–China, the decision by the United Kingdom to leave the European Union ("Brexit"), and the global New Coronavirus spread. In this background, a new normal is about to take hold, reflecting a continuous moderate growth of the global economy. This trend will significantly influence all the attached subsystems or sectors in the maritime transportation system, including infrastructure requirements, ship carrying capacity needs, ship design and technology, port developments and performance, and so on.

The primary impact of the slowing-down economy puts on the demand of maritime transport. In 2017–2019, the international maritime trade shares similar moderate growths with the global economy. According to the "Review of Maritime Transport 2019" by UNCTAD [1], although the global maritime trade reaches a new milestone of 11 billion tons in 2019, the growth is only 2.7%, not only lower than 4.1% in 2017, but also lower than 3.0% average from 1970 to 2017 [1]. Figure 1.1 and Table 1.1 respectively show the total cargo volumes of specific types in ton-miles and tons.

The moderate growth of maritime transport demand shall introduce more competition between different players, i.e., shipowners or port administrators and other stakeholders. This trend may re-shape the market structure since many less-efficient sectors in the maritime transportation system will fall into the brutal struggle between embracing the technology evolutions or being eliminated.

© The Author(s) 2021
S. Fang and H. Wang, *Optimization-Based Energy Management for Multi-energy Maritime Grids*, Springer Series on Naval Architecture, Marine Engineering, Shipbuilding and Shipping 11, https://doi.org/10.1007/978-981-33-6734-0_1

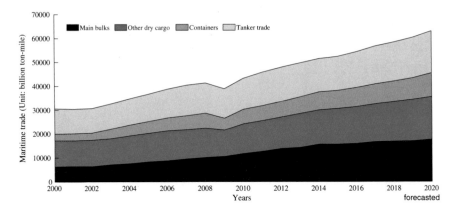

Fig. 1.1 Development of international maritime trade (Unit: billion ton-miles), reprinted from [1], open access

Table 1.1 Specific types of international maritime trade (Unit: million ton) reprinted from [1], open access

Year	Tanker trade	Main bulk	Other dry cargo	Total (all cargos)
2000	2163	1186	2635	5984
2005	2422	1579	3108	7109
2006	2698	1676	3328	7702
2007	2747	1811	3478	7702
2008	2742	1911	3578	8231
2009	2641	1998	3218	7857
2010	2752	2232	3423	8408
2011	2785	2364	3626	8775
2012	2840	2564	3791	9195
2013	2828	2734	3951	9513
2014	2825	2964	4054	9842
2015	2932	2930	4161	10,023
2016	3058	3009	4228	10,295
2017	3146	3151	4419	10,716
2018	3194	3210	4601	11,005

1.1.2   Ship Supply Capacity and Market Structure

The new trend of moderate growth also defines the recent supply-side development of the maritime transportation. In 2019, the world's commercial fleet consists of 95,402 ships, with a combined tonnage of 1.97 billion dwt [1]. The share of each principal vessel type is shown in the following Table 1.2.

Table 1.2 World fleet by principal vessel type (Unit: dwt), reprinted from [1], open access

Principal types	2018	2019	Increments
Oil tankers	562,035 29.2%	567,533 28.7%	0.98%
Bulk carriers	818,921 42.5%	842,438 42.6%	2.87%
General cargo ships	73,951 3.8%	74,000 3.7%	0.07%
Containers	253,275 13.1%	265,668 13.4%	4.89%
Other types	218,002 11.3%	226,854 11.5%	4.06%
Gas carriers	64,407 3.3%	69,078 3.5%	7.25%
Chemical tanker	44,457 2.3%	46,297 2.3%	4.14%
Offshore vessels	78,629 4.1%	80,453 4.1%	2.79%
Passenger vessels	6922 0.4%	7097 0.4%	2.53%
Other	23,946 1.2%	23,929 1.2%	−0.07
World total	**1,926,183**	**1,976,491**	**2.61**

From the Table 1.2, the oversupply of ship capacity remains a prominent characteristic for most of shipping sectors. Among all sectors, the gas carriers experience the highest growth rate at 7.25%, which is driven by the significant expansion of the liquefied natural gas (LNG) trade [2]. Then the container fleet follows at 5% increment. On the contrary, the chemical-tanker and dry-bulk-carrier segments both only experience moderate growths, and the oil tanker segment even suffers a downward trend.

In summary, the oversupply of the ship capacity will further reduce the average freight fare and cut down the profits. Some new technologies are therefore motivated to integrate into the maritime transportation system to gain efficiency improvement and competitional advantage.

1.1.3 Shipping Services and Ports

One effect of the market re-shaping is to enlarge the average sizes of ships since megaships generally have cheaper transportation costs than smaller ships. This trend is suggested in Table 1.3 by the increasing of average vessel size in recent years.

The increasing trend of vessel size has great impacts on the port terminals, as well as the shipyards and the inland logistics. The resulted-in influences come from

Table 1.3 Vessel size distribution to service years (Unit: dwt), reprinted from [1], open access

Types	Service years				
	0–4	5–9	10–14	15–19	20+
Bulk carriers	81,482	77,757	71,592	64,156	52,622
Container ships	83,362	66,050	43,565	38,031	19,579
General cargo	8770	7507	5255	6360	2725
Oil tanker	82,577	78,314	73,092	90,578	8241
Others	10,461	6548	8839	8136	4214
All ships	44,370	39,985	30,696	30,946	6342

two aspects: (1) mega-ships generally have limited access to many ports since draft restrictions or berth-length requirements, which makes the mega-ships can only call for services in some ports; (2) larger ships normally call at fewer ports than smaller ships during one voyage, and less calls with greater cargo volumes will create greater pressure on the operation of ports.

From above, as ships become larger, the ports and terminals that can accommodate the service-calls become limited, which means the main ports around the world, such as Singapore, Shanghai, Istanbul, Houston, Genoa, Hamburg, and so on, will face more competitions and challenges in the future. New equipment and technologies are required for the future large ports to efficiently provide at least three types of services to the berthed-in ships.

(1) Logistic services, including loading/unloading cargos from the onboard to the stacking areas, the restacking of cargos in the stackyard, the transportation to the inland logistic systems, and so on. This type of service is conventional but in current situations, large ports are required to further enhance the cargo handling efficiency and reduce the dwell-time in berth to strengthen their competitiveness.
(2) Electrical services, namely the on-shore power supply, or cold-ironing technology. For the future efficient ports, cold-ironing technology is necessary since it can greatly reduce the gas emission of the berthed-in ships in the harbor territory. According to [3], cold-ironing technology will become a mandatory requirement for large ports in the future.
(3) Heating/cooling services. For specialized cargo such as refrigerated goods, large ports need providing reefer slots, and for future cruise ships, large ports may also provide on-shore heating/cooling services to the onboard passengers [3].

To efficiently and economically provide the above services, the future ports are need to be significantly upgraded in both infrastructure planning and management framework. It should be noted that the ports not only include the mainland ones, but also include those in islands, or "general ports" in various ocean platforms, i.e., ocean oilfields, ocean wind farms, or drilling platforms.

1.1.4 The Path to the Green Shipping

Besides the above motivations, the entry into force of several global environmental policies and the adoption of some voluntary standards also have some fundamental impacts on the maritime transportation system and set the following two main targets to achieve the future green shipping.

(1) Relieving the heavy reliance on oil for propulsion.

Generally, more than 50% of the oil demand around the world is concentrated in the transport sector [4], and the global oil demand for maritime transportation is more than 300 million tonnes and accounts for 86% of the transport sector in 2012 [5]. According to [6], more than half of the fuel consumption increment in transportation is from maritime usage before 2040 if no further actions.

Additionally, the oil used for maritime transportation often has lower quality than other types of oil in the transport sector, i.e., denser and higher carbon-hydrogen ratio, as well as having more "polluted elements". For example, the IFO380 is a frequently used oil type for large container ships, which has more than 3.8% sulfur, much higher than the light-oil used in land-based transportation.

As a result, the great consumption and low quality of maritime oil make the maritime transportation system emit diversified gas emissions, and the heavy reliance on oil for propulsion, therefore, becomes the main obstacle to limit the development of green shipping. The research on alternative fuels or energy sources should raise global concerns, such as hydrogen fuel and ammonia fuel, fuel cell technologies, and energy storage [4, 7–9].

(2) Reducing greenhouse and polluting gas emissions.

The gas emission of maritime transportation usually has three types: carbon dioxide, sulfoxide, and nitrogen oxide. The carbon dioxide is regarded as the main culprit for the greenhouse effect and has been raised as a global concern since the subscription of the Kyoto Protocol in 1997 [10]. As for the sulfoxide and nitrogen oxide, they are viewed as two main types of polluted gas emissions, which are responsible for the acid rains and the ozone hole, respectively [11].

For global sustainable development, those three types of gases are all under strict surveillance, and for the future green shipping, multiple policies have been raised to address different types of gas emissions.

For the carbon dioxide, the International Maritime Organization (IMO) has announced an ambitious target to reduce 70% greenhouse gas (GHG) emission in 2050 compared with 2008 [12], shown as Fig. 1.2.

For better controlling the minimum required level of energy efficiency, the Energy Efficiency Design Index (EEDI) and Energy Efficiency Operating Index (EEOI) were established as the IMO's strategies. Specifically, EEDI is a ship designing index proposed by the Marine Environment Protection Committee (MEPC 62) in 2011 [13]. Then in MEPC 63, four guidelines are amended in MARPOL Annex VI to

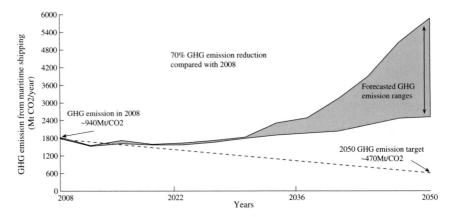

Fig. 1.2 Reduction target of GHG emission from maritime transportation, reprinted from [12], open access

further implement the EEDI as a mandatory regulation [14–17]. As for the EEOI, it is recommended by the Ship Energy Efficiency Management Plan (SEEMP) to manage the efficiency performance of ships and fleets over time using [18]. Both of EEDI and EEOI have been adopted by various ship companies.

For the sulfoxide, IMO has set certain limits since the year of 2000, shown as Fig. 1.3. From 1st, January 2020, the "ever strictest sulfur limit in history" has entered into force for the compulsory usage of low-sulfur fuel (0.5%), or the gas scrubber integration, or the alternative fuels.

The first sulfur limit was introduced in the revised MARPOL Annex VI (Prevention of Air Pollution from Ships) and the concept of designated sulfur emission control area (SECA) was created correspondingly [12]. The Baltic Sea, North Sea,

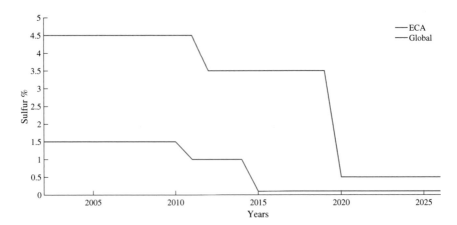

Fig. 1.3 Fuel sulfur limits in ECA and globe, reprinted from [12], open access

Table 1.4 NOx emission limits (MARPOL Annex VI)

Tier	Year	NOx limit (g/kWh)		
		Rated engine speed (r/min)		
		<130	130–2000	>2000
I	2000	17	$45 \cdot n^{-0.2}$	9.8
II	2011	14.4	$45 \cdot n^{-0.23}$	7.7
III	2016	3.4	$45 \cdot n^{-0.2}$	1.96

and North America have been designated as SECAs since 1997, 2005, and 2010, respectively. In 2011, the Caribbean Sea of United States has been designated as SECA. In October 2016, the regulation was confirmed at the MEPC 70, which dictates that from 2020 onward, the global limit of sulfur content will be 0.50% (outside SECAs), referring to the "ever strictest sulfur limit" [19].

For the nitrogen oxide, MARPOL has set tier I–III emission standards based on the speed of main engines and the ship-ages, shown as Table 1.4 [12]. Tier I is for the old ships built before 2000, and tier-II is the current NOx limit standard, and tier III is for the ships built after 2016 and sailing in nitrogen ECAs (NECAs). When outside the NECAs, the ships should follow tier II. It should be noted that the SECAs in the United States have been already set as NECAs. For the European SECAs (North and Baltic Sea), the NOx limits will be enforced from 2021.

Accordingly, to fulfill the above ambitious targets on gas emission reduction, considerable investments should be going into the research and development for new technologies, such as better hydrodynamics in ships, more energy-efficient engines, efficient ships with new configurations, lower carbon or carbon-free fuels, the renewable integrations and more advanced energy management systems.

1.2 Promising Technologies

1.2.1 Overview

To achieve future green and efficient shipping, many technologies have been already or about to implement in maritime transportation. They are mainly classified as (1) the technical designs and (2) the alternative fuel or energy sources. The details are shown in the following Table 1.5.

In the following context, the promising technologies related to the electrification of maritime transportation (in the bold context above) are illustrated in detail to show their usages.

Table 1.5 Promising technologies for the future green shipping

Technical designs for energy efficiency improvement	Alternative fuel or energy sources
Light construction material	**Hydrogen fuel cells**
Ship-hull optimization	Hydrogen as fuel
Propulsion-improvement devices	**Shipboard energy storage**
Air lubrication system	**Ammonia fuel cells**
Ballast-water system design	Ammonia as fuel
Engine and auxiliary system	Synthetic methane
Energy-efficiency measures	Synthetic diesel
All-electric ship	**Renewable power integration**
Gas capture system	Advanced biofuels
Multi-energy management	

1.2.2 Selected Technical Designs for Energy Efficiency Improvement

1.2.2.1 All-Electric Ship (AES)

Long before gaining global concerns, the emergence of AES is quite early. In 1922, the first aircraft carrier of the United States named as "Langly (CV-1)" was converted from a coal carrier named as "Jupiter", shown in Fig. 1.4a.

This ship uses the configuration of "steamer-generator-electric machine" to drive the propeller and can be viewed as an embryo of AES. However, due to the technical limits at that time, the reliability of the power network in "Langly" was much lower than other similar mechanically-driven ships. Therefore, after the technical breakthrough in large-scale gearbox, the configuration of "Langly" has been hung on, and the mechanically-driven ships, which directly drive the propeller by the prime mover, have come to their golden age and dominate the configuration designs of ships until now and even the near future.

(a) Langly (CV-1)   (b) Zumwalt-class destroyer   (c) America-class amphibious assault ship

Fig. 1.4 Main representatives of all-electric ships

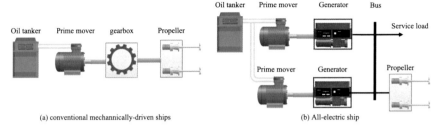

Fig. 1.5 Development of all-electric ships, reprinted from [20], with permission from IEEE

In the last decades, the advances in electrical engineering represented by the power electronic technologies have greatly improved the reliability of power systems, which progressively promote the development of microgrids. Nowadays, various types of microgrids have been utilized in different scenarios and applications. The bottleneck of AES has therefore been relieved.

With this great development, AES has gained the concerns from the shipping industry once again since higher energy efficiency and better controling ability. Currently, the configuration of AES has been already applied in warships, such as the Zumwalt-class destroyer and the America-class amphibious assault ship, which are shown in Fig. 1.4b and c, respectively.

The main advantage of AESs compared with conventional mechanically-driven ships is the usage of an "integrated power system" to dispatch energy, which can be shown as Fig. 1.5 [20].

In Fig. 1.5a, the propellers of conventional ships are directly driven by the prime mover via a gearbox. This configuration limits the speed of prime mover and therefore limits the energy efficiency improvement. Additionally, another system of "prime mover-generator-service load" is necessary for the mechanically-driven ships to supply power to the onboard electrical equipment, which leads to great unnecessary redundancy.

In an AES (Fig. 1.5b), electricity is the only secondary energy form onboard. All the shipboard loads, including the propulsion load and various types of service loads, are supplied by the "integrated power system". The energy flow can be precisely controlled to achieve the optimal energy efficiency, and the energy supply can be from multiple sources to improve the system reliability.

Due to the advantages above, AES has raised global concerns in recent years and has been viewed as the future direction of ship designs. Nowadays, this configuration begins to expand from the military applications to the commercial applications, such as the "ampere" ferry from Denmark [21], "puffer" cargo ship from China [22], and "Viking lady" off-shore support vessel (OSV) [23] and so on, which are shown in Fig. 1.6, respectively.

(a) "Ampere" ferry   (b) "Puffer" cargo ship   (c) "Viking lady" off-shore support vessel

Fig. 1.6 Some commercial all-electric ships

1.2.2.2 Cold-Ironing Technology

The propulsion systems of most ships consist of the main engines and the auxiliary engines, and when berthed in a port, the main engines will be kept off and the auxiliary engines are used to support the onboard load demand, such as lighting, refrigeration, kitchen, entertainment and so on. The auxiliary engines burn fuel to generate electricity and emit various types of gas emissions in the harbor territory, such as CO_2, SOx, and NOx, which brings a great amount of pollution.

Cold-ironing technology, or on-shore power supply, or shoreside power, is to supply the onboard hoteling load for the berthed-in ships by the port-side electricity, and the auxiliary engines onboard are all kept off, shown as Fig. 1.7 [24]. The electricity can be from the main grid, or port-side renewables and other clean fuels [24]. In the future, the cold-ironing technology will become a mandatory service from ports similar to the conventional logistic services.

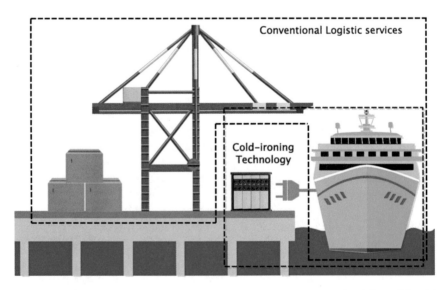

Fig. 1.7 Cold-ironing technology and conventional logistic services, reprinted from [24], with permission from IEEE

The first benefit of cold-ironing technology is the reduction of harbor gas emission. It is reported that the global harbor gas emission can have a 10% reduction by the integration of cold-ironing technology [25]. In UK ports, the cold-ironing technology can reduce 2% SOx emission. According to [26], cold-ironing technology reduces more than 57% of harbor gas emission in the Kaohsiung port in Taiwan. Secondly, cold-ironing technology may bring economic benefits to both the shipowners and the port authorities. Kenan et al. [27] shows in the regions of which the electricity price are lower than 0.19USD/kWh, the cold-ironing may reduce the operating cost of the berthed-in ships. In [25], the cold-ironing brings extra profits for the port with higher average handling time.

Cold-ironing technology is very suitable for the cruise ports, since when berthed-in, the cruise ships require a huge amount of power since many passengers staying on board [28, 29]. According to [30], an average of 29.3% of GHG emissions can be reduced in three different regional cruise ship cases when using cold-ironing. In other regions, the cruise ship ports can reduce 99.5% (Oslo, Norway), 85% (France) GHG emission by the cold-ironing technology, respectively.

Although the above outstanding advantages, the expansion of cold-ironing technology is still a challenging task. The main barriers include power quality [28], system stability [28], reliability and security [3], and synchronization problems [24]. Tsekouras and Kanellos [31] used a port-side reserved generator to improve the power quality of cold-ironing, and [32] proposed smart electrical interfaces to improve the performance of the cold-ironing facility. In [24], the synchronization problem of cold-ironing was investigated, and a control strategy is proposed to mitigate the voltage fluctuation when the ship plugged into the cold-ironing state, which is demonstrated by an OPAL-RT experiment.

1.2.2.3 The Electrification of Ports

The ports are need to provide adequate logistic services to the berthed-in ships by many different types of equipment. The main equipment includes quay crane (QC), rail-mounted gantry crane (RMG), rubber-tire gantry crane (RTG), reach stacker (RC), straddle carrier (SC), and lift trunk (LT), which are shown in the following Fig. 1.8.

QC is used for loading/unloading cargo or containers from the ship-side. RMG and RTG are used to stack containers in the stackyard, and the main difference is that the RMG moves on the rail and the RTG moves on rubber tires. RS is used to reach a container in the stackyard. SC and LT are used to transport the containers within the stackyard. Conventionally, the above equipment are almost manually-driven, and in recent years, highly automated port equipment, such as automated RTG, RMG, LT, SC, begins in usage to improve the efficiency and reduce labor usage [33]. The energy sources of those equipment also become diversified. Table 1.6 gives the possible energy sources of the above equipment.

From the Table 1.6, diesel and LNG are commonly used fuel types in port-side operation, which can power various port-side equipment. In addition from above,

(a) quay crane   (b) rail-mounted gantry   (c) rubber tire gantry

(d) reach stacker   (e) straddle carrier   (f) lift truck

Fig. 1.8 Main port-side logistic equipment

Table 1.6 Energy sources of different port-side equipment (data from [34])

	QC	RMG	RTG	RS	SC	LT
Diesel	Yes	No	Yes	Yes	Yes	Yes
Electricity	Yes	Yes	Yes	Yes	Yes	Yes
LNG	No	No	Yes	Yes	Yes	No

electricity is the most general energy source and can power all the main port-side equipment, and is also energy-efficient, easy to control, and convenient to fulfill automation, which makes the electrification of large ports as an irreversible trend in both shore-side operation and yard-side operation.

During the shore-side operation, the QCs can recover tremendous energy from the hoist-down movement [35]. In this way, the electrification and the integration of energy storage can shift the peak load of QCs and improve energy efficiency. In [35, 36], the peak load of QC can be reduced from 1211 to 330 kW with a supercapacitor integration. In [37], the peak load of QC is reduced from 1500 to 150 kW by the integration of energy storage. The shift of peak load not only represents higher energy efficiency but also can mitigate the influences on the port-side power system.

In yard-side operation, RMGs generally have higher energy efficiency than conventional RTGs since it is electrically-driven, but the advantage of RTG is the higher operating flexibility since its operation is not limited to the rails. In this sense, the electrification of RTG (E-RTG) can combine both advantages on energy efficiency and operating flexibility, which makes it a hot topic now and has reported gaining an 86.6% reduction in energy costs and 67% on GHG emission reduction

Table 1.7 Energy consumption comparison between RTG and E-RMG (data from [34])

	Energy consumption	Energy cost/year (k USD)	GHG emission (kg/container)
RTG	2.21 L/move	64048	5.96
E-RTG	3.02 kWh/move	8621	1.92

[38]. The energy consumption comparison between RTG and E-RTG is shown in Table 1.7, and the results clearly show the energy saving ability of E-RTG. The energy cost of E-RTG is only 13% of RTG, and the GHG emission is only 1/3 of RTG. Similarly, for other yard-side equipment, such as RS, SC, and LT, the hybrid diesel-electric engine system has already been integrated. In literature, the hybrid SC has gained 27.1% fuel efficiency improvement, and the traveling motion, hoisting motion and lowering motion consume 52, 31 and 11% less energy [39]. As above, with the development of electrical engineering technologies, especially the energy storage technology, all-electric RS, SC, and LT will soon become reality and achieve the zero-emission target.

1.2.2.4 Multi-energy Management

In recent years, with the development of global cold-chain supply, the refrigeration power demand grows very fast. In various studies, the energy consumption of refrigeration energy is now between 20 and 45% of the total energy consumption of ports [38, 40]. This suggests the need to improve the energy efficiency of reefer areas, such as determining the number of reefers, locations, and power plans. Additionally, due to the large scale of heating/cooling power demand on board, future cruise ships may also require heating/cooling power from the port-side. In summary, the above refrigeration power demand and the onboard heating/cooling power demand are both supplied by the heating/cooling flow, and can be viewed as "temperature-controled power demand" [40].

With the integration of heating/cooling flows, there will be at least three energy flows coordinated to each other in maritime grids, i.e., fossil fuel, electricity, and heating/cooling power, which makes the future maritime grid as multi-energy systems (MESs), and proper multi-energy management is essential for this special MESs.

Multi-energy management is a newly proposed management framework to coordinate multiple energy flows, shown in Fig. 1.9 [41–43], which shows different energy forms can convert to each other in MES to shift the peak load to fill up the valley, thus gaining higher energy efficiency compared with the single-energy system, such as the conventional power system. This management framework has been used in many land-based applications. In Jiangsu and Guangdong provinces of China, there are already system-scale projects of MESs.

In an MES, the main power sources are the upper electric network (UEN) and the upper gas network (UGN). The main power demands are the electricity demand, gas demand, heating demand, and cooling demand, which are supplied by the electrical

energy flow

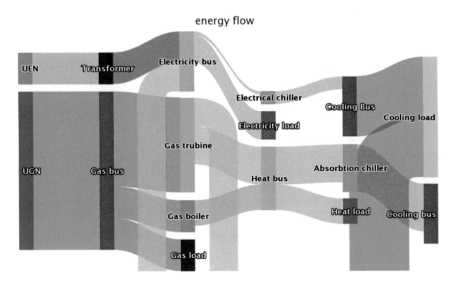

Fig. 1.9 Energy flows in a conventional multi-energy system

bus, gas bus, heating bus, and cooling bus, respectively. Among different energy forms, the gas turbine can generate electricity by burning gas. The by-produced heat can supply the heat load or the cooling load after the absorption chiller. The electricity also can be converted to gas by the power to gas equipment (P2G). In the future, the municipal water supply may also implement into MES since the expansion of the electric water pump.

However, the ships and ports have quite different operating scenarios compared with conventional land-based applications, such as extra electrical and logistic constraints, which makes current multi-energy management methods cannot be directly used, and further research efforts should be put into this field.

1.2.2.5   Gas Capture Systems

In Sect. 1.4, the main targets of the gas emission control have been discussed. The alternative fuel and electrification technologies are generally viewed as promising routes to resolve this energy efficiency problem. However, before the maturity of above technologies, the integration of gas capture system can be viewed as an effective transitional approach. With its integration, the gas emission can be captured and stored in a location and permanently away from the atmosphere, thus the energy efficiency (gas emission per unit task) can be improved with continuously using the conventional fossil fuel. Nowadays, the capture systems of CO_2, SOx, NOx are all mature technologies and ready to integrate into maritime grids.

Generally, the gas capture systems have three main working frameworks, which are shown in Fig. 1.10a–c, i.e., the pre-combustion, oxygen-fuel and post-combustion methods [44].

Among the above three working frameworks, the post-combustion method is the most frequently used (Fig. 1.10c) since a relatively simpler process, and the gas capture systems manufactured by the Wasilla and Mann are mostly using the framework of post-combustion.

In recent years, driven by the "ever strictest sulfur limit" planned to enter into force from 1st, January 2020 [19], many shipowners have planned to invest the shipboard gas capture systems to act as the transitional approach. With the gas capture system installed, the ships can continue to sail on the heavy oil (IFO380, 3.8% sulfur) with a lower price (cheaper than MGO, 0.5% sulfur) meanwhile meeting the environmental requirements. The initial cost-benefit analysis shows this investment can be refunded in 4–5 years under current oil prices [45, 46].

A typical illustration of gas capture system into ships is shown in Fig. 1.11 [47]. The emitted gas from the main and auxiliary engines are first absorbed and then stored in a storage. With sufficient energy supply, the gas capture system can reduce more than 70% gas emission [44].

However, most of the ships are not designed with the gas capture system, thus the onboard engines may not have enough capacities to supply the power demand after the installation, and this is one of the main obstacle to limit the gas capture system integration in the views of energy management. In [44], an extra generator is invested to supply the power demand of gas capture system, and in [47], the energy storage and onboard generators are coordinated to meet the power demand of the gas capture system, and the capturing rate is more than 90%.

1.2.3 Selected Alternative Fuels or Energy Sources

1.2.3.1 Renewable Power Generation

Generally, the renewable power generation integration into ships and ports is the fundamental approach to resolve the energy efficiency problem of maritime grids, i.e., when the penetration rate of renewable power generation increases, the usage of conventional fossil fuel will reduce. Until now, the integration of renewable power generation into maritime grids already has many practical cases, shown as Fig. 1.12a–c.

Figure 1.12a shows the "Zhongyuan Tengfei" photovoltaic (PV) integration project in 2016. The total PV module has 143.1 kW capacity and can provide power for lighting in 12 decks [48]. Figure 1.12b shows a conceptual hybrid renewable energy ship proposed by Sauter Carbon Offset company, Germany [49] in 2010. This ship uses wind and PV energy to sustain 16 knot speed with zero-emission. Figure 1.13c is the "Shangde Guosheng" ferry in Shanghai Expo, 2010, which has a length of 31.85 m, a breath of 9.8 meters, and a height of 7 m. Now, this ship serves as a tourist ship in Huangpu River, Shanghai [50].

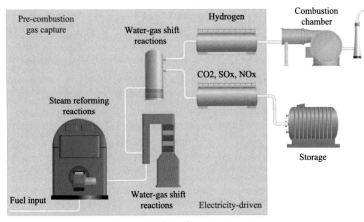

(a) Pre-combustion

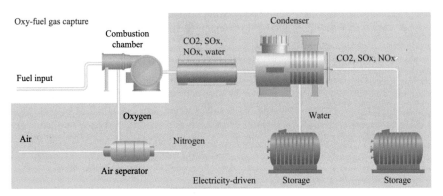

(b) Oxygen-fuel

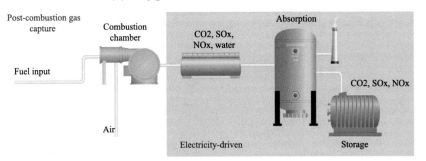

(c) Post-combustion

Fig. 1.10 Main gas capture working frameworks

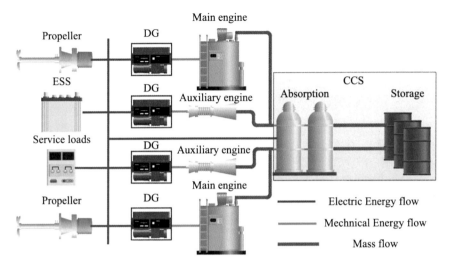

Fig. 1.11 Gas capture system into ships, reprinted from [47], with permission from IEEE

(a) "Zhongyuan Tengfei" shipboard photovoltaic project

(b) "Black magic" hybrid renewable concept ship   (c) "Sangde Guosheng" photovoltaic integrated ferry

Fig. 1.12 Practical renewable integrated ships

However, since the low energy density and limited installment area, renewable energy integration into ships can only supply a small part of total energy demand now, and due to the uncertainties, the shipboard energy management system needs to be upgraded to mitigate their influences [51].

On the port-side, the capacity of renewable energy is much higher since larger installment area. In the Jurong port (Singapore), the installed PV can generate more

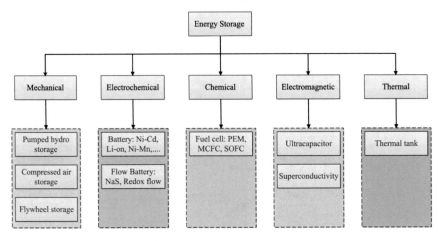

Fig. 1.13 Classification of energy storage

than 12 million kWh electricity per year [3]. In Houston port (US), the Spilman's island is planned for PV modules, and the potential PV capacity can be more than 4 MW [52]. In Hamburg port (Germany), the installed wind turbines scale up to 25.4 MW [53]. The above cases have demonstrated that the renewable power generation has a very promising future in the port-side applications.

1.2.3.2 Energy Storage and Fuel Cell Integration

Generally, electrical energy can be converted to many different forms for storage, which are shown as following Fig. 1.13 [54]. Among all the technologies in Fig. 1.13, pumped hydro storage, compressed air storage and flow battery are not suitable for maritime applications since the limits in locations and operating conditions. The superconductivity storage has very limited energy capacity, which is also not practical nowadays. The most promising energy storage technologies currently in maritime applications include flywheel, battery, ultra-capacitor, and thermal tank. Strictly speaking, fuel cell is a power source technology rather than a type of energy storage technology. But it has similar characteristics and operating conditions with conventional energy storage systems, i.e., no combustion process, small installment area, directly outputting electricity, no spinning components. In this section, the fuel cell is discussed together with the other conventional energy storage systems (ESSs).

In maritime applications, ESSs are used for (1) peak load shifting by the high energy density ESS; and (2) resolving power quality issues by the high power density ESS. In long-term timescale, the peak load shifting can mitigate the burdens of main power sources (generators) and the energy efficiency can be improved [55–62]. Then the high power density ESS can respond to the load fluctuations in short-term timescale to resolve the power quality problems [63–65].

Since no combustion process, fuel cells have higher generation efficiency and smaller unit-sizes than the traditional internal combustion engines, which is the promising power source technology for maritime applications in the future. At present, the hydrogen fuel cell based on proton exchange membradune technology is a relatively mature technology and has been used in the energy supply of submarine [66], but the production and storage of hydrogen are still expensive, which limits the further commercial applications of hydrogen fuel cells. On the other side, liquefied natural gas (LNG) and Maritime Diesel Oil (MDO) are currently the main fuel types in commercial ships, and the corresponding Molten Carbonate Fuel Cells (MCFC) and Solid Oxide Fuel Cells (SOFC) on these two types of fuels, therefore, have higher commercial values.

Currently, fuel cell acts as an auxiliary power source in ships, an illustration in all-electric ships can be shown as Fig. 1.14 [67].

In the Fig. 1.14, fuel cell is installed at one bus to supply the low-voltage hotel load and the propulsion load. Besides, there are two cases in Fig. 1.14, and the first case is the "Viking lady" offshore supporting vessel (OSV) has already installed a 330 kW fuel cell compared with the total generation system of 8040 kW [23]. The other case is, in 2019, the 712 ship institute of China has invented a 500 kW shipboard fuel cell.

All around the world, the fuel cell applications in ships are shown in Table 1.8, which includes both military and commercial applications. With the development of fuel cells, the capacity of the fuel cell will further increase to replace the current onboard spinning prime mover. However, there is still a long way to go before the fully replacement of internal combustion engine to fuel cell. Many obstacles, such as the energy management problems, the lifetime management problems, are still pending.

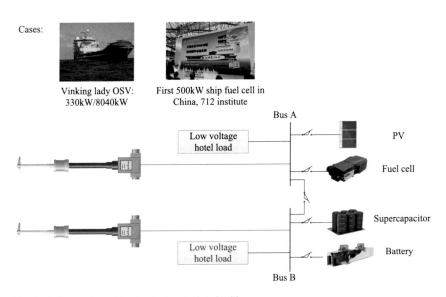

Cases:

Vinking lady OSV: 330kW/8040kW

First 500kW ship fuel cell in China, 712 institute

Bus A

Low voltage hotel load

PV

Fuel cell

Supercapacitor

Low voltage hotel load

Battery

Bus B

Fig. 1.14 Integration of fuel cell into all-electric ships

Table 1.8 Projects of some selected fuel cell based ships

Ship	Power	Fuel	References
Viking Lady	330 kW	LNG	[23]
SF-Breeze	100 kW	Hydrogen	[68]
PA-X-ELL	30 kW	Methanol	[69]
MV Undine	250 kW	Methanol	[70]
US SSFC	2.5 MW	Diesel	[71]
MC-WAP	500 kW	Diesel	[72]
MS Forester	100 kW	Diesel	[73]
212 submarine U31	330 kW	Hydrogen/Methanol	[74]
212 submarine U32	240 kW	Hydrogen/Methanol	[75]
S-80 Submarine	300 kW	Ethanol	[76]

1.2.3.3 Low-Carbon and Carbon-Free Fuel

Current maritime transportation highly relies on fossil oil for propulsion, especially the usage of heavy oil (IFO380). The replacement of heavy oil to low-carbon and even carbon-free fuel is the fundamental approach to resolve the energy efficiency problem of maritime transportation. Generally, alternative fuels are classified as low-carbon fuels, i.e., LNG, methanol, ethanol, and carbon-free fuels, i.e., hydrogen, ammonia.

Among the low-carbon fuels, LNG is the alternative fuel for heavy oil which gained the highest concerns [4]. It is estimated that the shipping sector can reduce more than 30% GHG emission by using LNG [4]. In 2018, the LNG transport ships have expanded by 7.25%, which represents the prosperous development of this sector [1]. Among the carbon-free fuels, hydrogen has been regarded as the future green fuel for a long time, since the only product after burning is water. The ship using hydrogen for propulsion can achieve zero gas emission [4]. Ammonia is a newly proposed carbon-free fuel in recent years, with the products of nitrogen and water [7].

Despite of the above great achievements, there are still many limits on the application of low-carbon and carbon-free fuels. The gaps come from the following four aspects, and a promising and practical alternative fuel should resolve all the below problems before it can replace the heavy oil for propulsion.

(1) Mass production

To become a practical alternative fuel, the primary problem is the mass production problem. Although Hydrogen, Methanol, and Ethanol are all vital raw materials in chemical industry, the global production of those fuels is still hard to sustain global shipping [7]. Among current fuels, LNG and ammonia are only two fuels with enough production capacity to sustain global shipping. LNG is for its great amount of natural resource reserves, and the ammonia is for its great production ability as a widely used chemical fertilizer [7].

(2) Commercial power source technologies

Current engines in ships are mostly designed for heavy oil. When changing to other fuels, the burning chambers should be reformulated or re-designed to keep the burning stability of fuels. However, except for the LNG, there still lacks large-scale commercial power source technologies for other alternative fuels to sustain long-distance of navigation [7]. Furthermore, some alternative fuels, like ammonia, are very hard to burn in conventional conditions. In this way, the fuel cell can be a very promising way for maritime applications since it has no burning process.

(3) Mass storage

The heavy oil usually has much higher energy density and less volatileness than the low-carbon and carbon-free fuels. When changing to alternative fuels, the storage conditions need to be adjusted, i.e., larger volume to sustain the navigation distance, proper sealing conditions to reduce the volatilization of fuels, and so on. For example, the volatilization loss of LNG transport in a week is about 5% by current technology. In this way, the mass storage technology of LNG, such as re-liquefaction, is needed urgently in this field.

(4) Global supply chain

Since the shipping fuel is used all around the world, the transport cost will greatly influence the expansions and usages of alternative fuel, which means the candidate alternative fuels should have a mature and complete global supply chain to reduce its transport cost. In fact, LNG and ammonia are the only two fuels with a complete and mature global supply chain [7].

1.3 Next-Generation Maritime Grids

From the above illustrations, the main characteristic of the next-generation maritime grids is the trend of electrification and the involvement of multiple energy flows, and current green shipping technologies are convenient to implement into the maritime grids. With various new technologies integrated, the next-generation maritime grids are defined as those local energy networks (combined with electrical, fossil fuel and heating/cooling energy networks) installed in harbor ports, ships, ferries, or vessels, which consists of generation, storage and critical loads, and can operate either in grid-connected or in islanded modes and operate under both the constraints of power system and maritime transportation system. In the following context, two main representatives of the next-generation maritime grids, i.e., shipboard microgrid, seaport microgrid, are illustrated and then the coordination between them is shown.

1.3.1 Shipboard Microgrid

With full electrification, the integrated power system of ships formulates a microgrid, and the ship is referred as "AES", which is illustrated in Fig. 1.15.

From above, the shipboard microgrid consists of both an energy network (blue lines and arrows) and a communication network (green lines and arrows). The generators and battery deliver power via the energy network to meet the propulsion and service loads. The propulsion load is used to drive the ship. The service load supplies electricity to various onboard equipment, including the onboard radar, navigation system, air conditioning, as well as the gas capture system in Sect. 1.2.2.5. In the future, fuel cells may further replace the generators to act as the main power sources. To further improve the energy efficiency of AES, renewable energy can be integrated, like the photovoltaic modules in Fig. 1.15. As for the communication network, the shipboard energy management system (EMS) can optimally calculate the generators and battery outputs and then send the dispatch signals to each component by it.

A special case is the cruise ship since the large scale of thermal load demand onboard. In fact, the shipboard microgrid of cruise ship can be viewed as an MES [77], shown as Fig. 1.16.

The main differences between Figs. 1.15 and 1.16 are the involvement of heat flow. In a cruise ship, the combined cooling/heating power generator (CCHP) and power to cooling/heating (PTC) equipment are installed to act as the heating sources, and the electrical flow and heating flow should be coordinated to achieve a better economic and environmental behaviors.

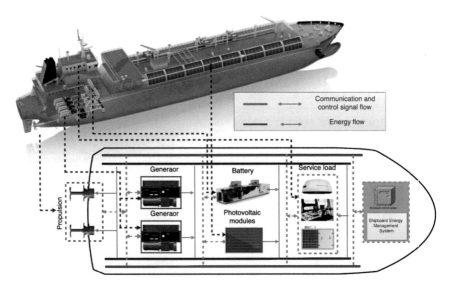

Fig. 1.15 Typical topology of future all-electric ships, reprinted from [24], with permission from IEEE

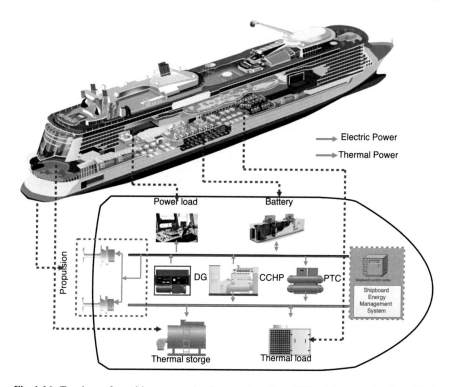

Fig. 1.16 Topology of a multi-energy cruise ship, reprinted from [77], with permission from IEEE

1.3.2 Seaport Microgrid

Seaport microgrid is a newly proposed concept for seaport management, which is depicted by [24]. The incentive of the seaport microgrid is to make it as an energy district to improve renewable energy penetration and enhance the grid storage capacity by selling the electricity to the market through the main grid. There are already many practical cases around the world of seaport microgrid. In [78], the author advocated the harbor area as a unique territory that should have new business models with its energy plan. In [79], two practical projects of seaport microgrids in Hamburg (German) and Genoa (Italy), are manifested in detail, and the operating data proves the validity of seaport microgrid.

A typical seaport microgrid is illustrated in Fig. 1.17. Generally, the seaport is connected with the main grid and various renewable energy are integrated, i.e., seaport wind farms and PV farms. All the port-side equipment, including the quay cranes, gantry cranes, transferring trunks, are electrical-driven.

The seaport provides four types of services to the berthed-in ships: (1) logistic service. The berth allocation and quay crane scheduling for loading/unloading cargo; (2) fuel transportation. Unloading or refilling fuel for the berthed-in ships; (3) cold-ironing. Providing electricity to the berthed-in ships; and (4) refrigeration reefer for

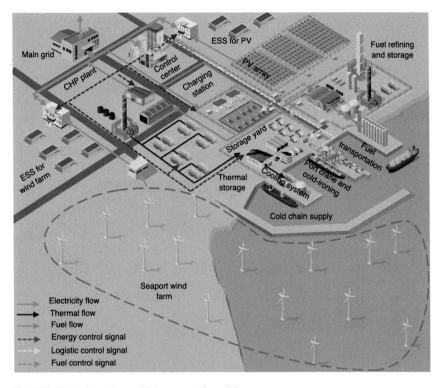

Fig. 1.17 Typical topology of future port microgrid

the cold-chain supply. With the above multiple energy flows involved, the future seaport microgrid is a "**maritime multi-energy system**" [24], and the port central control should give both the energy and logistic control signals to each sub-system in the seaport [24].

1.3.3 Coordination Between Shipboard and Seaport Microgrids

In the future, the connection between the ship and the port is no longer limited in the logistic-side, and will be also expanded to the electrical-side. Figure 1.18 shows the coordination between the ships and ports. When the ship berthed in, the seaport will allocate a berth position and some corresponding port cranes for loading/unloading onboard cargos. In the electric-side, the berthed-in ship is directly connected to the seaport by the AC/DC converters, and all the load demands are met by the on-shore side. The seaport becomes a coordinated electric-logistic multi-microgrid system.

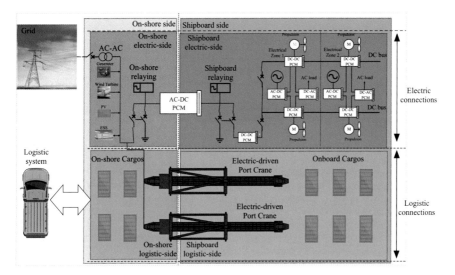

Fig. 1.18 Coordination between the seaport microgrid and shipboard microgrid, reprinted from [24], with permission from IEEE

1.4 Summary

In this chapter, we have concluded that the maritime grids are those local energy networks installed in harbor ports, ships, ferries, or vessels, which consists of generation, storage and critical loads, and are able to operate either in grid-connected or in islanded modes and operate under both the constraints of power system and maritime transportation system. After the illustration of various promising technologies which are about to integrate, the implementation of next-generation maritime grids is suggested to be a promising approach to resolve the energy efficiency problem of the maritime transportation system, and may have the ability to re-shape the future relationship between the ocean and inland.

Using full-electrification as the backbone, the future maritime grids, i.e., all-electric ships, seaport microgrids, and various electrified ocean platforms, become the "**maritime multi-energy system**", which requires an advanced energy management system to achieve the economic and environmental targets. From the aspect of electrical engineering, the future maritime grids are a special type of power systems. In land-based applications, the optimization-based power system operation has been extensively studied and should be expanded to the maritime grids for future usages. This is also the main goal of this book, i.e., the optimization-based energy management for the next-generation maritime grids.

References

1. United Nations Conference on Trade and Development (UNCTD): Review of Maritime Transport 2019, UNCTAD/RMT/2019/Corr.1
2. Ng, A.K.Y., Monios, J., Jiang, C.: Maritime Transport and Regional Sustainability. Elsevier (2020)
3. Iris, Ç., Lam, J.: A review of energy efficiency in ports: operational strategies, technologies and energy management systems. Renew. Sustain. Energy Rev. **112**, 170–182 (2019)
4. Psaraftis, H.N.: Sustainable Shipping: A Cross-Disciplinary View. Springer (2019)
5. IMO: Third IMO GHG study 2014. Executive summary and final report. International Maritime Organization, London
6. EIA: International energy outlook 2017. U.S. Energy Information Administration, Norway
7. Nick, A., Tim, S.: Sailing on Solar: Could Green Ammonia Decarbonise International Shipping? Environmental Defense Fund, London (2019)
8. Zou, G.: Integrated Energy Solutions to Smart And Green Shipping: 2019 Edition, VTT Technical Research Centre of Finland Ltd.
9. ICCT: Reducing greenhouse gas emissions from ships cost effectiveness of available options. White Paper, International Council for Clean Transportation (2011)
10. IMO: Amendments to the annex of the protocol of 1997 to amend the international convention for the prevention of pollution from ships, 1973, as modified by the protocol of 1978 relating thereto, annex 3, resolution mepc. 278(70) (MEPC 70/18/Add.1)
11. Ullinane, K., Cullinane, S.: Atmospheric emissions from shipping: the need for regulation and approaches to compliance. Transp. Rev. **33**(4), 377–401
12. IMO: Revised MARPOL Annex VI: Regulations for the prevention of air pollution from ships. Report Marine Environment Protection Committee (2008)
13. IMO: Resolution MEPC.203(62), Amendments to the Annex of the Protocol of 1997 to amend the International Convention for the Prevention of Pollution from Ships, 1973, as modified by the Protocol of 1978 (2011)
14. IMO: Resolution MEPC.212(63), 2012 Guidelines on the Method of Calculation of the Attained EEDI for new ships (2012a)
15. IMO: Resolution MEPC.213(63), 2012 Guidelines for the development of a ship energy efficiency management plan (SEEMP) (2012b)
16. IMO: Resolution MEPC.214(63), 2012 Guidelines on Survey and Certification of the EEDI (2012c)
17. IMO: Resolution MEPC.215(63), Guidelines for Calculation of Reference Lines for Use With the Energy Efficiency Design Index (EEDI) (2012d)
18. IMO: Resolution MEPC.282(70), Guidelines for the Development of a Ship Energy Efficiency Management Plan (SEEMP) (2016)
19. Cullinane, K., Bergqvist, R.: Emission control areas and their impact on maritime transport. Transp. Res. Part D: Transp. Environ. **28**, 1–5 (2014)
20. Fang, S., Gou, B., et al.: Optimal hierarchical management of shipboard multi-battery energy storage system using a data-driven degradation model. IEEE Trans. Transp. Electrif. **5**(4), 1306–1318 (2019)
21. Norway electric ferry cuts emissions by 95%, costs by 80%. https://reneweconomy.com.au/norway-electric-ferry-cuts-emissions-95-costs-80-65811/
22. A new all-electric cargo ship with a massive 2.4 MWh battery pack launches in China. https://electrek.co/2017/12/04/all-electric-cargo-ship-battery-china/
23. Lady, V. http://maritimeinteriorpoland.com/references/viking-lady/. Accessed 27 August 2018
24. Fang, S., Wang, Y., et al.: Toward future green maritime transportation: an overview of seaport microgrids and all-electric ships. IEEE Trans. Veh. Technol. **69**(1), 207–220 (2020)
25. Thalis, Z., Jacob, N., et al.: Evaluation of cold ironing and speed reduction policies to reduce ship emissions near and at ports. Maritime Econ. Logist. **16**(4), 371–398 (2014)
26. Chang, C., Wang, C.: Evaluating the effects of green port policy: case study of Kaohsiung harbor in Taiwan. Transp. Res. Part D: Transp. Environ. **17**(3), 185–189 (2012)

27. Kenan, Y., Görkem, K., et al.: Energy cost assessment of shoreside power supply considering the smart grid concept: a case study for a bulk carrier ship. Maritime Policy Manag. **8839**(January), 1–14 (2016)
28. Tseng, P., Pilcher, N.: A study of the potential of shore power for the port of Kaohsiung, Taiwan: to introduce or not to introduce? Res. Transp. Bus. Manag. **17**, 83–91 (2015)
29. Ballini, F., Bozzo, R.: Air pollution from ships in ports: the socio-economic benefit of cold-ironing technology. Res. Transp. Bus. Manag. **17**, 92–98 (2015)
30. William, J.: Assessment of CO_2 and priority pollutant reduction by installation of shoreside power. Resourc. Conserv. Recycle **54**(7), 462–467 (2010)
31. Tsekouras, G.J., Kanellos, F.D.: Ship to shore connection—reliability analysis of ship power system. In: 2016 XXII International Conference On Electrical Machines (ICEM), September 2016, pp. 2955–2961
32. Coppola, T., Fantauzzi, M., et al.: A sustainable electrical interface to mitigate emissions due to power supply in ports. Renew. Sustain. Energy Rev. **54**, 816–823 (2016)
33. Hossein, G., Debjit, R., René, K.: Sea container terminals: new technologies and OR models. Maritime Econ. Logist. **18**(2), 103–140 (2016)
34. Gordon, W., Thomas, S.: Energy consumption and container terminal efficiency. FAL Bull **350**(6), 1–10 (2016)
35. Zhao, N., Schofield, N., et al.: Hybrid power-train for port crane energy recovery. In: IEEE Conference and Expo Transportation Electrification Asia-Pacific (ITEC Asia-Pacific), pp. 1–6, August 2014
36. Tran, T.: Study of electrical usage and demand at the container terminal, Ph.D. Thesis, Deakin University, July 2012
37. Parise, G., Honorati, A.: Port cranes with energy balanced drive. In: AEIT Annual Conference: From Research to Industry: The Need for a More Effective Technology Transfer (AEIT), pp. 1–5, September 2014
38. Greencranes: Green technologies and eco-efficient alternatives for cranes and operations at port container terminals. GREENCRANES project. Technical report October 2012
39. Hangga, P., Shinoda, T.: Motion-based energy analysis methodology for hybrid straddle carrier towards eco-friendly container handling system. J. Eastern Asia Soc. Transp. Study **11**, 2412–2431 (2015)
40. Acciaro, M., Wilmsmeier, G.: Energy efficiency in maritime logistics chains. Res. Transp. Bus. Manag. **17**, 1–7 (2015)
41. Tasdighi, M., Ghasemi, H., et al.: Residential microgrid scheduling based on smart meters data and temperature dependent thermal load modeling. IEEE Trans. Smart Grid **5**(1), 349–357 (2014)
42. Liu, W., Wen, F., Xue, Y.: Power-to-gas technology in energy systems: current status and prospects of potential operation strategies. J. Modern Power Syst. Clean Energy **5**(3), 439–450 (2017)
43. Mohammadi, M., Noorollahi, Y., Mohammadi-ivatloo, B., et al.: Energy hub: from a model to a concept: a review. Renew. Sustain. Energy Rev. **80**, 1512–1527 (2017)
44. Luo, X., Wang, M.: Study of solvent-based carbon capture for cargo ships through process modelling and simulation. Appl. Energy **95**, 402–413 (2017)
45. PSE Ltd.: DNV and PSE report on ship carbon capture & storage. https://www.psenterprise. com/news/news-pressreleases-dnv-pse-ccs-report. Accessed 6 September 2016
46. Ship-technology: Onboard carbon capture: dream or reality? http://www.ship-technology.com/ features/featureonboard-carbon-capture-dreamor-reality/. Accessed 25 September 2016
47. Fang, S., Xu, Y., et al.: Optimal sizing of shipboard carbon capture system for maritime greenhouse emission control. IEEE Trans. Ind. Appl. **55**(6), 5543–5553 (2019)
48. The biggest photovoltaic integrated ship: Zhongyuan Tengfei. https://www.sohu.com/a/694 98034_160309
49. Zero emission renewable power ship by Sauter Carbon Offset. http://www.kumotor.com/art icle_page.php?owt01=20110117170304886
50. Shangde Guosheng tourist ship. http://beidiao.w84.mc-test.com/Info/commerce/id/439

51. Fang, S., Xu, Y., et al.: Data-driven robust coordination of generation and demand-side in photovoltaic integrated all-electric ship microgrids. IEEE Trans. Power Syst. (2019) (In press)
52. Molavia, A., Shib, J., et al.: Enabling smart ports through the integration of microgrids: a two-stage stochastic programming approach. Appl. Energy **258**, 114022 (2020)
53. HPA: Energy cooperation, port of Hamburg Technical Report, August 2015
54. Muhammad, M., Yacine, T., et al.: Energy storage systems for shipboard microgrids: a review. Energies **11**, 3492 (2018)
55. Kanellos, F.D.: Optimal power management with GHG emissions limitation in all-electric ship power systems comprising energy storage systems. IEEE Trans. Power Syst. **29**(1), 330–339 (2014)
56. Kanellos, F.D., Tsekouras, G.J., Hatziargyriou, N.D.: Optimal demand-side management and power generation scheduling in an all-electric ship. IEEE Trans. Sustain. Energy **5**(4), 1166–1175 (2014)
57. Shang, C., Srinivasan, D., Reindl, T.: Economic and environmental generation and voyage scheduling of all-electric ships. IEEE Trans. Power Syst. **31**(5), 4087–4096 (2016)
58. Hou, J., Sun, J., et al.: Mitigating power fluctuations in electric ship propulsion with hybrid energy storage system: design and analysis. IEEE J. Oceanic Eng. **43**(1), 93–107 (2018)
59. Fang, S., Xu, Y., Li, Z., et al.: Two-step multi-objective management of hybrid energy storage system in all-electric ship microgrids. IEEE Trans. Veh. Technol. **68**(4), 3361–3372 (2019)
60. Boveri, A., Silvestro, F., Molinas, M., et al.: Optimal sizing of energy storage systems for shipboard applications. IEEE Trans. Energy Conv. Early Access (2018)
61. Lan, H., Wen, S., et al.: Optimal sizing of hybrid PV/diesel/battery in ship power system. Appl. Energy **158**, 26–34 (2015)
62. Wen, S., Lan, H., et al.: Allocation of ESS by interval optimization method considering impact of ship swinging on hybrid PV/diesel ship power system. Appl. Energy **175**, 158–167 (2016)
63. Xie, C., Zhang, C.: Research on the ship electric propulsion system network power quality with flywheel energy storage. In: Proceedings of the Power and Energy Engineering Conference (APPEEC) Asia-Pacific, Chengdu, China, pp. 1–3, 28–31 March 2010
64. Samineni, S., Johnson, B., et al.: Modeling and analysis of a flywheel energy storage system for voltage sag correction. IEEE Trans. Indus. Appl. **42**, 42–52 (2006)
65. Mo, R., Li, H.: Hybrid energy storage system with active filter function for shipboard MVDC system applications based on isolated modular multilevel DC/DC converter. IEEE J. Emerg. Sel. Top. Power Electron. **5**, 79–87 (2017)
66. Ellis, M., Spakovsky, M.R., Nelson, D.J.: Fuel cell systems: efficient, flexible energy conversion for the 21st century. Proc. IEEE **89**(12), 1808–1818 (2001)
67. Alireza, T., Ali, N., et al.: Fuel cell power management using genetic expression programming in all-electric ships. IEEE Trans. Energy Convers. **32**(2), 779–787 (2017)
68. SF-BREEZE. https://energy.sandia.gov/transportation-energy/hydrogen/markettransformation/maritime-fuel-cells/sf-breeze/. Accessed 27 August 2018
69. Pa-x-ell. http://www.e4ships.de/aims-35.html. Accessed 27 August 2018
70. METHAPU Prototypes Methanol SOFC for Ships. Fuel Cells Bull. **5**, 4–5. 2859(08)70190-1 (2008)
71. SFC Fuel Cells for US Army, Major Order from German Military. Fuel Cells Bull. **6**, 4 (2012)
72. Jafarzadeh, S., Schjølberg, I.: Emission reduction in shipping using hydrogen and fuel cells. In: Proceedings of the ASME International Conference on Ocean, Offshore and Arctic Engineering, Trondheim, Norway, p. V010T09A011, 25–30 June 2017
73. MS Forester. https://shipandbunker.com/news/emea/914341-fuel-cell-technologysuccessfully-tested-on-two-vessels. Accessed 27 August 2018
74. A Class Submarine. http://www.seaforces.org/marint/German-Navy/Submarine/Type-212A-class.htm. Accessed 27 August 2018
75. SSK S-80 Class Submarine. https://www.naval-technology.com/projects/ssk-s-80-classsubmarine/. Accessed 27 August 2018
76. Kumm, W.H., Lisie, H.L.: Feasibility study of repowering the USCGC VINDICATOR (WMEC-3) with modular diesel fueled direct fuel cells. Arctic Energies Ltd Severna Park MD, Groton, MA, USA (1997)

77. Fang, S., Fang, Y., et al.: Optimal heterogeneous energy storage management for multi-energy cruise ships. IEEE Syst. J. (2020) (In press)
78. Lamberti, T., Sorce, A., et al.: Smart port: exploiting renewable energy and storage potential of moored boats, Oceans-2015. Genova, Italy (2015)
79. Acciaro, M., Ghiara, H., et al.: Energy management in seaports: a new role for port authorities. Energy Policy **71**(14), 4–12 (2014)

Chapter 2
Basics for Optimization Problem

2.1 Overview of Optimization Problems

2.1.1 General Forms

In different engineering scenarios, the maximizing or minimizing of some functions relative to some sets are common problems. The corresponding set often represents a range of choices available in a certain situation, and the "solution" infers the "best" or "optimal" choices in this scenario. Some common applications include "minimal cost, maximal profit, minimal error, optimal design, and optimal management". This type of problem has a general form as follows.

$$\min_{x} / \max_{x} f(x)$$
$$s.t. h(x) \leq 0 \tag{2.1}$$
$$g(x) = 0, \forall x \in S$$

In (2.1), $f(x)$ is the objective function, and represents the management tasks; "min" and "max" represent the minimizing and maximizing of $f(x)$, respectively; x is the decision variables, and represents the choices of administrator; $h(x) \leq 0$ and $g(x) = 0$ are the inequality and equality constraints to limit the decision variables, which represents the limitations on the choices of administrator by different operating scenarios; S is the original set for the decision variables, such as continuous variables, binary variables, integer variables, and so on. In this problem, the model expects to find the "best" or "optimal" solution "x^*" which meets the minimization or maximization of $f(x)$, and in reality, this may represent the minimization of costs or the maximization of profits. Here we give a simple case, Example 2.1, for the optimization problems.

© The Author(s) 2021
S. Fang and H. Wang, *Optimization-Based Energy Management for Multi-energy Maritime Grids*, Springer Series on Naval Architecture, Marine Engineering, Shipbuilding and Shipping 11, https://doi.org/10.1007/978-981-33-6734-0_2

Example 2.1: Knapsack Problem
Assuming we have n types of goods and indexed by $i \in 1, 2, 3 \ldots, n$. Each good values W_i and the size is S_i, and we have a knapsack with the capacity of C. The problem is how we can pack the goods with the highest value? The model of this problem is shown as follows.

$$\max \sum_{i=1}^{n} W_i \cdot x_i$$
$$s.t. \ \sum_{i=1}^{n} S_i \cdot x_i \leq C \tag{2.2}$$
$$x_i \in \{0, 1\}$$

In (2.2), x_i is the "decision variables", and represents the choice of the ith good or not. If choosing the ith good into the knapsack, $x_i = 1$ and if not, $x_i = 0$. $x_i \in \{0, 1\}$ is the original set of "decision variables". $\sum_{i=1}^{n} W_i \cdot x_i$ is the "objective function", and represents the total values of the selected goods, and $\sum_{i=1}^{n} S_i \cdot x_i \leq C$ is the "constraint", represents the total sizes of goods that should be smaller or equal to the capacity of the knapsack. The "best" or "optimal" solution "$\{x_i^*, i \in 1, 2, 3 \ldots, n\}$" can achieve the maximization of $\sum_{i=1}^{n} W_i \cdot x_i$.

Case Study for Example 2.1
Here we test a simple case, and the parameters are shown as follows: $n = 5$, $\{W_i | i = 1, 2, \ldots, 5\} = [2, 3, 1, 4, 7]$, and $\{S_i | i = 1, 2, \ldots, 5\} = [2, 2, 1, 2, 3]$, and $C = 6$. The simulation results are shown in Fig. 2.1.

From the Fig. 2.1, the best solution for Example 2.1 is to select the 3rd, 4th and 5th goods, and the maximal total value is 12, and the total size of goods is 6.

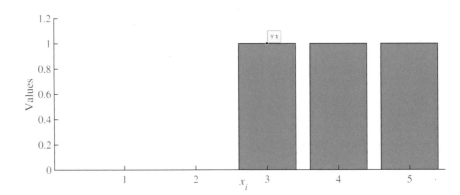

Fig. 2.1 Simulation results of Example 2.1

2.1.2 Classifications of Optimization Problems

(1) Classifications by decision variables

The decision variable x may consist of different number types, such as continuous variables, binary variables, and integer variables, and the combination of different number types produces different optimization problems. For example, if x only consists of continuous variables, then problem (2.1) is "continuous optimization", and if x only consists of binary variables or integer variables, then problem (2.1) is "binary optimization" or "integer optimization". If x simultaneously has continuous variables and integer variables, then problem (2.1) is "mixed-integer optimization".

(2) Classifications by the objective function

Generally, objective function $f(x)$ can be a scalar or vector and based on it, problem (2.1) is "single-objective optimization" when $f(x)$ is a scalar and "multi-objective optimization" or "vector optimization" when $f(x)$ is a vector.

(3) Linear optimization and non-linear optimization

In practical cases, $f(x), h(x)$ and $g(x)$ may have different mathematical characteristics. If $f(x), h(x)$ and $g(x)$ are all linear functions, problem (2.1) is "linear optimization (LP)", and is "non-linear optimization (NLP)" if anyone in $f(x), h(x)$ and $g(x)$ is non-linear. Specifically, if $f(x)$ is non-linear, and $h(x)$ and $g(x)$ are both linear, then problem (2.1) is "linear constrained and non-linear objective optimization (LCNLP)", and if $f(x)$ is linear, and $h(x)$ and $g(x)$ are both non-linear, then problem (2.1) is "non-linear constrained and linear objective optimization". Here we give some typical cases, if $f(x), h(x)$ and $g(x)$ are all polynomials and the largest power is two, then problem (2.1) is "quadratic optimization (QP)". Similarly, we can define the "quadratic-objective quadratic-constrained optimization (QCQP)", "quadratic-objective linear-constrained optimization (LCQP)", and so on.

(4) Convex optimization and non-convex optimization

Before introducing the convex optimization and non-convex optimization, the convex function and convex set should be described in the first place. Firstly, convex functions should meet (2.3) for any x_1 and x_2 in the domain of $f(x)$ [1].

$$f(\alpha \cdot x_1 + (1 - \alpha) \cdot x_2) \leq \alpha \cdot f(x_1) + (1 - \alpha) \cdot f(x_2), \forall \alpha \in [0, 1] \qquad (2.3)$$

Then the convex set S should meet: for any two points in S, denoted as s_1 and s_2, their linear combination $\alpha \cdot s_1 + (1 - \alpha) \cdot s_2$ is still within S [1]. Illustrations for convex function and convex set are shown in Fig. 2.2a and b, respectively.

From Fig. 2.2a, $x_3 = \alpha \cdot x_1 + (1 - \alpha) \cdot x_2$ and $f_3(x_3) < \alpha \cdot f_3(x_1) + (1 - \alpha) \cdot f_3(x_2)$, thus $f_3(x)$ is a convex function. Similarly, $f_1(x)$ is a concave function, and $f_2(x)$ is a convex function and also a concave function. From Fig. 2.2b, any linear combination

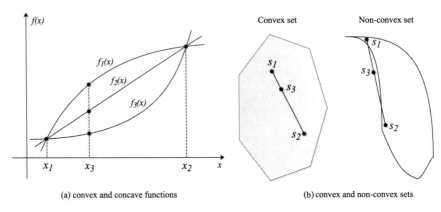

(a) convex and concave functions   (b) convex and non-convex sets

Fig. 2.2   Illustrations of convex functions and convex sets

of s_1 and s_2 belongs to the same set, which represents the convexity. For the non-convex set, at least one combination of s_1 and s_2 is outside the same set, shown in Fig. 2.2b.

With the above definitions, (2.1) is a convex optimization problem when the following two conditions satisfied: (1) $f(x)$ is convex in case of minimization and concave in case of maximization; (2) $S = \{x | h(x) \leq 0, g(x) = 0, \forall x \in S\}$ is a convex set. The main characteristic of the convex optimization compared with non-convex optimization is, a local optimal solution of the convex optimization is also the global optimal solution of this convex optimization [1]. This characteristic greatly benefits the applications of convex optimization, and in reality, if we can model or reformulate the problems as convex optimization, then the global optimal solution can be obtained after resolving any local optimal ones. This is one of the main reasons for "the main watershed in optimization problem is not between the linear ones and non-linear ones, but the convex ones and non-convex ones" [1].

In summary, the classification methods can be combined to characterize different optimization problems, such as "mixed-integer linear optimization (MILP)", "mixed-integer non-linear optimization (MINLP)", "mixed-integer quadratic optimization (MIQCP)", and so on.

2.2   Optimization Problems with Uncertainties

Uncertainties are inevitable in reality since the measurement and control both have errors. To ensure safety and reliability, considering uncertainties in optimization problems is necessary, and stochastic optimization, robust optimization, and interval optimization are three main types.

2.2.1 Stochastic Optimization

A general form of stochastic optimization is shown as (2.4) [2].

$$\min_{x \in X} g(x) + E\left(\min_{y \in Y(x,\xi)} f(y) \right) \qquad (2.4)$$

In stochastic optimization (Eq. 2.4), x is the first stage decision variables which are not determined by uncertainties; X is the feasible region of x; $g(x)$ is the objective function of the first stage; ξ is the uncertain variables, and $Y(x, \xi)$ is the feasible region of y determined by x and ξ; $f(y)$ is the objective function of the second stage; $E(\cdot)$ is the expectation. In this model, the uncertain variable ξ is depicted by the probability distribution, such as the probability distribution of equipment failure, or the probability distribution of renewable energy output, and so on. Then stochastic optimization seeks the optimal solution within the feasible region defined by the probability distributions. To clearly show the stochastic optimization, Example 2.2 is reformulated as follows.

Example 2.2: Stochastic Knapsack Problem
Based on all the assumptions of Example 2.1, we further assume that for $\forall i \in \{1, 2, \ldots, n_f\}$, W_i is a constant, and for $\forall i \in \{n_f, n_f + 1, \ldots, n\}$, $W_i = W_c + \Delta W_i$, where W_c is a constant and ΔW_i follows a pre-given distribution ψ. Then the original knapsack problem becomes (2.5).

$$\max \left(\sum_{i=1}^{n_f} W_i \cdot x_i + \sum_{i=n_f}^{n} W_c \cdot x_i \right) + E\left(\max\left(\sum_{i=n_f}^{n} \Delta W_i \cdot x_i \right) \right)$$
$$s.t. \ \sum_{i=1}^{n} S_i \cdot x_i \leq C \qquad (2.5)$$
$$x_i \in \{0, 1\}, \quad \Delta W_i \in \psi$$

where $\left(\sum_{i=1}^{n_f} W_i \cdot x_i + \sum_{i=n_f}^{n} W_c \cdot x_i \right)$ is "$-g(x)$", and the "$-$" is to transform the maximization of (2.5) to the minimization of (2.4), and this term is not influenced by the uncertainties; $\sum_{i=n_f}^{n} \Delta W_i \cdot x_i$ is $f(y)$ which is influenced by the uncertainties; and $x = \{x_i | i = 1, 2, \ldots, n_f\}$, and $y = \{x_i | i = n_f, n_f + 1, \ldots, n\}$.

Case Study for Example 2.2 Here we test a simple case, and the parameters are shown as follows: $n = 5$, $\{W_i | i = 1, 2, 3\} = [2, 3, 1]$, and $\{W_c | i = 4, 5\} = [4, 7]$, and $\{\Delta W_i | i = 4, 5\}$ is normally distributed as $\mathcal{N}(0, 1)$, and $\{S_i | i = 1, 2, \ldots, 5\} = [2, 2, 1, 2, 3]$, and $C = 6$. The simulation results are shown in Fig. 2.3a and b.

From the Fig. 2.3a, the best solution for Example 2.2 is also to select the 3rd, 4th and 5th goods, and the expected maximal total value is 12.23, and the total size of goods is 6. The main difference between the stochastic optimization (2.5) and the conventional deterministic problem (2.2) is the uncertainties of ΔW_i will cause the uncertainties of objective function, which is shown as Fig. 2.3 (b).

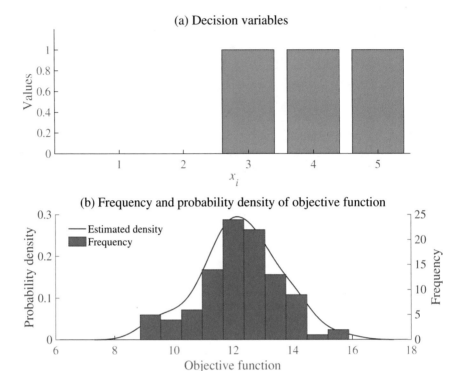

Fig. 2.3 Simulation results of Example 2.2

2.2.2 Robust Optimization

A general form of robust optimization is shown as (2.6) [3].

$$\min_{x \in X} g(x) + \max_{\xi \in U} \left(\min_{y \in Y(x,\xi)} f(y) \right) \tag{2.6}$$

In robust optimization (Eq. 2.6), the main difference is the uncertain variable ξ is described by the uncertainty set U, including the upper/lower limits and the uncertainty budget. Then robust optimization seeks the optimal solution in the worst case in the defined uncertainty set and therefore brings conservatism. With above, the primary problem of the uncertainty modeling is how to determine the feasible regions, such as the probability distributions in stochastic optimization and the uncertainty set in robust optimization. Similarly, we can give a robust knapsack problem as Example 2.3.

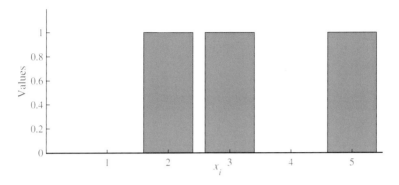

Fig. 2.4 Simulation results of Example 2.3

Example 2.3: Robust Knapsack Problem
Based on all the assumptions of Examples 2.1 and 2.2, we further assume ΔW_i is within a range denoted as $[W_L, W_U]$, and the robust knapsack problem can be shown as (2.7). The meaning of each part is similar to the stochastic model of (2.5).

$$\max\left(\sum_{i=1}^{n_f} W_i \cdot x_i + \sum_{i=n_f}^{n} W_c \cdot x_i\right) + \min_{\Delta W_i}\left(\max\left(\sum_{i=n_f}^{n} \Delta W_i \cdot x_i\right)\right)$$
$$s.t. \ \sum_{i=1}^{n} S_i \cdot x_i \leq C \tag{2.7}$$
$$x_i \in \{0, 1\}, \ \Delta W_i \in [W_L, W_U]$$

Case Study for Example 2.3
Here we test a simple case, and the parameters are shown as follows: $n = 5$, $\{W_i | i = 1, 2, 3\} = [2, 3, 1]$, and $\{W_c | i = 4, 5\} = [4, 7]$, and $\{\Delta W_i | i = 4, 5\} \in [-2, 1]$, and $\{S_i | i = 1, 2, \ldots, 5\} = [2, 2, 1, 2, 3]$, and $C = 6$. The simulation results are shown in Fig. 2.4.

From the Fig. 2.4, the best solution for Example 2.3 in robust optimization becomes the 2nd, 3rd and 5th goods, and the value of the objective function is 9. This change is due to the risk of the 4th good, since in the worst case, its value becomes 2, and it is not worthy to select. From the above results, we can find the results of robust optimization is conserve.

2.2.3 Interval Optimization

Interval optimization can be viewed as an enhancement of robust optimization and consisted of a lower sub-problem and an upper sub-problem, shown as (2.8) [4], and the upper sub-problem is similar with the robust optimization of (2.6). It should be noted that, for the maximization problem, the lower sub-problem is a robust optimization problem. The main advantage of interval optimization is the interval obtained can be used to analyze the influences of uncertainties on the system. A case is given as Example 2.4.

$$
\left[\underbrace{\min_{x \in X} g(x) + \min_{\xi \in U}\left(\min_{y \in Y(x,\xi)} f(y) \right)}_{Lower\ sub-problem}, \underbrace{\min_{x \in X} g(x) + \max_{\xi \in U}\left(\min_{y \in Y(x,\xi)} f(y) \right)}_{Upper\ sub-problem} \right] \tag{2.8}
$$

Example 2.4:Interval Knapsack Problem

$$
Lower:\max\left(\sum_{i=1}^{n_f} W_i \cdot x_i + \sum_{i=n_f}^{n} W_c \cdot x_i \right) + \min_{\Delta W_i}\left(\max\left(\sum_{i=n_f}^{n} \Delta W_i x_i \right) \right) \tag{2.9}
$$

$$
Upper:\max\left(\sum_{i=1}^{n_f} W_i \cdot x_i + \sum_{i=n_f}^{n} W_c \cdot x_i \right) + \max_{\Delta W_i}\left(\max\left(\sum_{i=n_f}^{n} \Delta W_i x_i \right) \right) \tag{2.10}
$$

Case Study for Example 2.4

The parameters of Example 2.4 is the same as Example 2.3, and the decision variables keep the same as Example 2.3, shown as Fig. 2.5, and the range of objective function is [9, 12]. From this, we can see the interval optimization can give both pessimistic and optimistic scenarios.

In summary, how to get the range of uncertain variables, i.e., the probability distribution function or the uncertainty set of ξ, is the basic problem of the optimization model. Nowadays, with the development of measurement and communication technology, more operating data can be transmitted and stored in the control center in

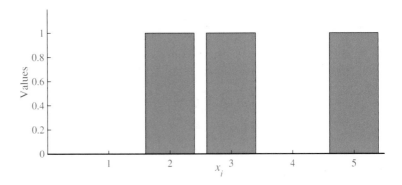

Fig. 2.5 Simulation results of Example 2.4

real-time. How to use this type of massive data to model the feasible region of uncertainty has become a hot topic, and various methods have been proposed. This topic will discuss in Chap. 4.

2.3 Convex Optimization

The importance of convex optimization has been emphasized in the former context, and in practical cases, we always want to model or reformulate a complex problem as convex ones, and the semi-definite programming (SDP) and the second-order cone programming (SOCP) are two classic types and have been well studied, which has gained the concerns from both academic and industry.

2.3.1 Semi-definite Programming

The general form of SDP is given as (2.11) [5].

$$\min A_0 \cdot X$$
$$s.t. \ A_p \cdot X = b_p, (p = 1, 2, .., m) \tag{2.11}$$
$$0 \preccurlyeq X \in S^n$$

where A_0, A_p are all coefficient matrixes; X is the decision matrix which should be semi-definite; b_p is a coefficient vector; S^n is the real space with n dimensions. Conventional linear optimization (LP) and quadratic optimization (QP) can be both formulated as SDP by defining $X = x \cdot x^T$ [6], then many commercial solvers can be used to solve the reformulated SDP for the global optimal solution, like Sedumi.

2.3.2 Second-Order Cone Programming

The general form of SOCP is given as (2.12) [7].

$$\min f^T \cdot x$$
$$s.t. \ ||A_i \cdot x + b_i||_2 \le c_i^T \cdot x + d_i, i = 1, 2, \ldots, n \tag{2.12}$$
$$F \cdot x = g$$

where f^T, A_i, b_i, c_i^T, d_i, F, g are all coefficient vectors or matrixes; x is the decision variables. It should be noted that the objective function is no need to be linear, and quadratic objective function can also be solved like conventional SOCP. Similarly, many types of optimization problems can be reformulated as SOCP, and several cases are given below to show the usages of SOCP.

(1)   Quadratic terms

For quadratic terms like x^2, it can be relaxed by the following (2.13) [8].

$$x^2 \leq W, ||2W, \gamma - x||_2 \leq \gamma + q \tag{2.13}$$

(2)   Bilinear terms

For bilinear terms like $x \cdot y$, it can be relaxed by the following (2.14) [8].

$$x \cdot y = z \tag{2.14}$$

$$\frac{1}{2}(x + y)^2 - \frac{1}{2}(x^2 + y^2) \leq z$$
$$\frac{1}{2}(x^2 + y^2) - \frac{1}{2}(x + y)^2 \leq z \tag{2.15}$$

In (2.15), $-\frac{1}{2}(x^2 + y^2)$ and $-\frac{1}{2}(x + y)^2$ are concave, and the following convex-concave procedure can be used to convexify them [9].

$$\frac{1}{2}(x + y)^2 - \frac{1}{2}(\bar{x}^2 + \bar{y}^2) - \bar{x} \cdot (x - \bar{x}) - \bar{y} \cdot (y - \bar{y}) \leq z$$
$$\frac{1}{2}(x^2 + y^2) - \frac{1}{2}(\bar{x} + \bar{y})^2 - (\bar{x} + \bar{y})(x - \bar{x} + y - \bar{y}) \leq z \tag{2.16}$$

where $(\bar{x}, \bar{y})$ is a constant reference point.

(3)   Exponential terms

For bilinear terms like e^x, it can be relaxed by the following (2.17).

$$y = e^x, \log(y) \geq x, \log(y) \leq x \tag{2.17}$$

Then at a reference point $\bar{y}$, (2.17) can be reformulated as (2.18) similarly by the convex-concave procedure [9].

$$\log(\bar{y}) + \frac{1}{\bar{y}} \cdot (y - \bar{y}) \leq x \tag{2.18}$$

2.4   Optimization Frameworks

2.4.1   Two-Stage Optimization

In reality, there are many cases that the decision variables cannot be determined in the same time, and this is the main motivation of two-stage optimization. The stochastic and robust optimization models in (2.4) and (2.6) are both two-stage optimization. Here we give a general form of two-stage optimization as (2.19) [10].

$$
\min_{x \in X} g(x) + \min_{y \in Y} f(y)
$$
$$
s.t. \ \ l(x) \leq 0, h(y) \leq 0 \tag{2.19}
$$
$$
G(x, y) \leq 0
$$

In the above formulation, $g(x)$ and $f(y)$ are the objective functions of the first stage and the second stage, respectively; and x, y are the corresponding decision variables; $l(x) \leq 0$, $h(y) \leq 0$ are the corresponding constraints and $G(x, y) \leq 0$ is the coupling constraints. It should be noted that, two-stage means x, y cannot be determined in the same time. To clarify this problem, Example 2.5 is given below.

Example 2.5: Two-stage Knapsack problem
Based on all the assumptions of Example 2.1, we assume that the ith good when $i = 1,2,...,n_1$ is available now and the ith good when $i = n_2, \ldots, n$ will be available after some times, and $n_2 \leq n_1$. The objective is still the maximization of the total values, but each good can only be selected one time. Then the optimization problem becomes (2.20).

$$
\min_{x \in X} \sum_{i=1}^{n_1} W_i \cdot x_i + \sum_{j=n_2}^{n} W_i \cdot y_j
$$
$$
s.t. \ \ \sum_{i=1}^{n_1} S_i \cdot x_i \leq C, \ \sum_{j=n_2}^{n} S_j \cdot y_j \leq C, x_i \in \{0, 1\}, y_j \in \{0, 1\} \tag{2.20}
$$
$$
x_i + y_j \leq 1, i \in n_1, \ldots, n_2
$$

In the above formulation, $\sum_{i=1}^{n_1} W_i \cdot x_i$ and $\sum_{j=n_2}^{n} W_i \cdot y_j$ are the objective functions of the first stage and the second stage, respectively, and $\sum_{i=1}^{n_1} S_i \cdot x_i \leq C, \sum_{j=n_2}^{n} S_j \cdot y_j \leq C$ are their corresponding constraints, and $x_i + y_j \leq 1, i \in n_1, \ldots, n_2$ is the coupling constraints.

Case study for Example 2.5
Here we test a simple case, and the parameters are shown as follows: $n = 5, n_1 = 3$ and $n_1 = 2$, and $\{W_i | i = 1, 2, \ldots, 5\} = [2, 3, 1, 4, 7]$, and $\{S_i | i = 1, 2, \ldots, 5\} = [2, 2, 1, 2, 3]$, and $C = 6$. The simulation results are shown in Fig. 2.6.
   From Fig. 2.6, the final objective function is 13 by the final selections of the 1st, 2nd, 3rd, and 5th goods. In the first stage, the capacity of the knapsack is 4, and

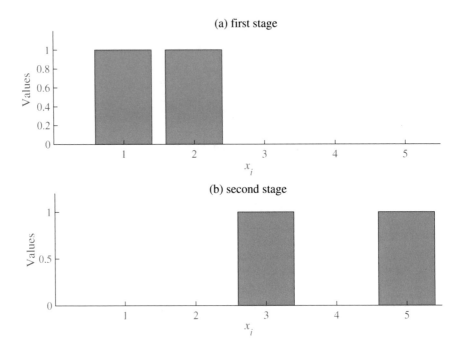

Fig. 2.6 Simulation results of Example 2.5

the 1st and 2nd goods are selected, then in the second stage, the 3rd and 5th goods are selected, and no good has been selected for twice. If the coupling constraint $x_i + y_j \leq 1, i \in n_1, \ldots, n_2$ is eliminated and the value of the 3rd good comes to 2, then the final selections are the 2nd good, and 3rd good for twice and the 5th good, and the objective function comes to 14.

In summary, the coupling constraint in two-stage optimization is essential which could influence the final results. Which is proved by many practical cases, the modifications on the coupling constraints benefit the objective function [10, 11].

2.4.2 Bi-level Optimization

Bi-level optimization is a special type of two-stage optimization and has a general formulation as following (2.21) [12]. In the following formulation, $F(x, y)$ represents the upper-level objective function and $f(x, z)$ represents the lower-level objective function. Similarly, x represents the upper-level decision vector and y represents the lower-level decision vector. $G_i(x, y)$ and $g(x, z)$ represents the inequality constraint functions at the upper and lower levels respectively. We can find that y is the decision variable of $F(x, y)$ and also the optimal decision variable to minimize $f(x, z)$. The upper and lower levels are coupled to achieve the overall optimum. Here we also give Example 2.6.

$$\min_{x \in X, y \in Y} F(x, y)$$
$$s.t. \ G_i(x, y) \leq 0, i \in 1, 2, \ldots, I \tag{2.21}$$
$$y \in arg \ \min_{z \in Y}\{f(x, z): g(x, z) \leq 0\}$$

Example 2.6: Bi-level Knapsack problem

Based on all the assumptions of Example 2.1, we assume inside the outer knapsack with the capacity of C_o, there is a small bag which holds the most valuable goods, and the capacity is C_s, and the objective is to maximize the total values and also in the small bag. Then the optimization problem becomes (2.22).

$$\max \left(\sum_{i=1}^{n_f} W_i \cdot x_i + \sum_{i=n_f}^{n} W_i \cdot x_i \right)$$
$$s.t. \ \sum_{i=1}^{n_f} S_i \cdot x_i \leq C_o - C_s, x_i \in \{0, 1\} \tag{2.22}$$
$$\{x_i | i = n_f, n_f + 1, \ldots, n\} = argmax \left\{ \sum_{i=n_f}^{n} W_i \cdot x_i \ | \ \sum_{i=n_f}^{n} S_i \cdot x_i \leq C_s \right\}$$

In the above formulation, $\sum_{i=1}^{n_f} W_i \cdot x_i + \sum_{i=n_f}^{n} W_i \cdot x_i$ and $\sum_{i=n_f}^{n} W_i \cdot x_i$ are the objective functions of the upper level and lower level, respectively. $\sum_{i=1}^{n_f} S_i \cdot x_i \leq C_o - C_s$ and $\sum_{i=n_f}^{n} S_i \cdot x_i \leq C_s$ are their constraints, respectively.

Case study for Example 2.6

Here we test a simple case, and the parameters are shown as follows: $n = 5, n_f = 2$, and $\{W_i | i = 1, 2, \ldots, 5\} = [2, 3, 1, 4, 7]$, and $\{S_i | i = 1, 2, \ldots, 5\} = [2, 2, 1, 2, 3]$, and $C_s = 3$, and $C_o = 6$. The simulation results are shown in Fig. 2.7. From the results, the lower level selects the 5th good in the first place and then in the upper level, the 2nd and 3rd goods are selected, and the final objective function comes to 11.

2.5 Summary

This chapter has briefly introduced the frequently used optimization models in engineering, and listed several important literature in the references for the readers. Simple testcases are also given to show different types of optimization models, and the models above will be used in the rest chapters of this book.

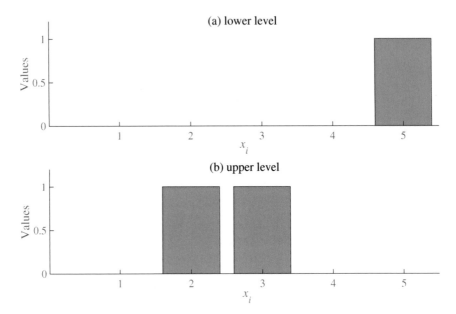

Fig. 2.7 Simulation results of Example 2.6

References

1. Boyd, S., Vandenberghe, L.: Convex Optimization. Cambridge University Press (2004)
2. Heyman, D., Sobel, J.: Stochastic Models in Operations Research: Stochastic Optimization. Courier Corporation (2004)
3. Ben-Tal, A., El Ghaoui, L., Nemirovski, A.: Robust Optimization. Princeton University Press (2009)
4. Bhurjee, A.K., Panda, G.: Efficient solution of interval optimization problem. Math. Methods Oper. Res. **76**(3), 273–288 (2012)
5. Vandenberghe, L., Boyd, S.: Semidefinite programming. SIAM Rev. **38**(1), 49–95 (1996)
6. Boyd, S., Vandenberghe, L.: Semidefinite programming relaxations of non-convex problems in control and combinatorial optimization. In: Communications, Computation, Control, and Signal Processing. Springer, Boston, MA, pp. 279–287 (1997)
7. Lobo, M., Vandenberghe, L., Boyd, S., et al.: Applications of second-order cone programming. Linear Aalgebra Appl. **284**(1–3), 193–228 (1998)
8. Zamzam, S., Dall'Anese, E., Zhao, C., et al.: Optimal water–power flow-problem: formulation and distributed optimal solution. IEEE Trans. Control Netw. Syst. **6**(1), 37–47 (2018)
9. Lipp, T., Boyd, S.: Variations and extension of the convex–concave procedure. Optim. Eng. **17**(2), 263–287 (2016)
10. Zeng, B., Zhao, L.: Solving two-stage robust optimization problems using a column-and-constraint generation method. Oper. Res. Lett. **41**(5), 457–461 (2013)
11. Zhao, C., Wang, J., Watson, J., et al.: Multi-stage robust unit commitment considering wind and demand response uncertainties. IEEE Trans. Power Syst. **28**(3), 2708–2717 (2013)
12. Dempe, S.: Foundations of Bilevel Programming. Springer Science & Business Media (2002)

Chapter 3
Mathematical Formulation of Management Targets

3.1 Overview of the Management Tasks

Generally, maritime grids are designed to fulfil different missions, and during the missions, different management tasks should be achieved, which can be mainly classified as five types: (1) thermodynamic tasks; (2) environmental tasks; (3) economic tasks; (4) logistic tasks; (5) service tasks. Five types are grouped in Table 3.1.

Chapter 1 has clarified the focus of this book: the long-term energy management of maritime grids, therefore the thermodynamic tasks are beyond the scope. In this chapter, six tasks are selected for their deep relationsip with the maritime grids.

3.2 Navigation Tasks

3.2.1 Typical Cases

As we all know, the ocean area covers more than 70 percent of our planet. In the following Fig. 3.1, we can see the main maritime shipping routes have connected the whole world. With the help of this meshed route grid and the main junctions, the bulk cargos, containers, and passengers are freely traveling by ships. Therefore, the primary management tasks for the maritime grids are the navigation tasks, which require the ships to arrive at the destination on time.

There are many different types of navigation tasks, and in this section, three representative cases are to show the navigation tasks for ships: (1) ferry routes for a short two-way trips; (2) cruise routes for a long-distance traveling; (3) cargo/container ship route for inland/oversea trading.

© The Author(s) 2021
S. Fang and H. Wang, *Optimization-Based Energy Management for Multi-energy Maritime Grids*, Springer Series on Naval Architecture, Marine Engineering, Shipbuilding and Shipping 11, https://doi.org/10.1007/978-981-33-6734-0_3

Table 3.1 Management tasks
and articles

Management tasks		Article
Thermodynamic	Exergetic efficiency	[1]
	Exergy destruction	[2]
	Net power output	[3–5]
	Power output	[6]
	Maximum temperature	[7]
	Voltage fluctuation	[8]
Environmental	EEOI	[9]
	Gas emission	[10]
Economic	Operation cost	[9, 10]
	Fuel consumption	[9, 10]
	Investment cost	[10]
Logistic tasks	On-time rate	[11]
	Weather routing	[12]
	Cargo handling	[13]
Service tasks	Customer satisfaction	[14]

3.2.1.1 Ferry Route

In many rivers, lakes, or straits, it is neither economic nor environmental to build bridges above the water, and in those areas, ferries can usually act as one of the main transport vehicles. In the following Fig. 3.2, two ferry routes are shown.

Figure 3.2a shows the ferry routes between Singapore to Batam. The current ferries connect two ports in Singapore with five ports in Batam. There is a combination of 100 ferry-crossings each day across seven ferry routes, and seven routes are operated by four ferry companies, including Sindo Ferry, Horizon Fast Ferry, Batam Fast Ferry, and Majestic Fast Ferry, with the shortest crossing taking around 50 min (HarbourFront Centre to Sekupang) [16].

Figure 3.2b shows the ferry route between the banks of the Yangtze River in Chongqing, China, which connects the downtown of Chongqing "Chaotianmen Square" with the Nanbin Road. This route is a famous tourism route in China which has very nice urban views of the downtown and therefore it is not suitable to build a bridge. This ferry line has been operated for more than 30 years and the entire voyage consumes about 30 min.

Besides the above two cases, ferries are widely used in many other places in the world. Especially in north Europe, ferries can convey both the passengers and cars to pass many strait georges which are not suitable to build bridges.

Generally, the voyage of ferries is usually much shorter than other ships like the cruises or cargo ships, and the ferry routes are often located near cities or towns. Therefore the ferries are the pioneers of electrified ships for environmental concerns.

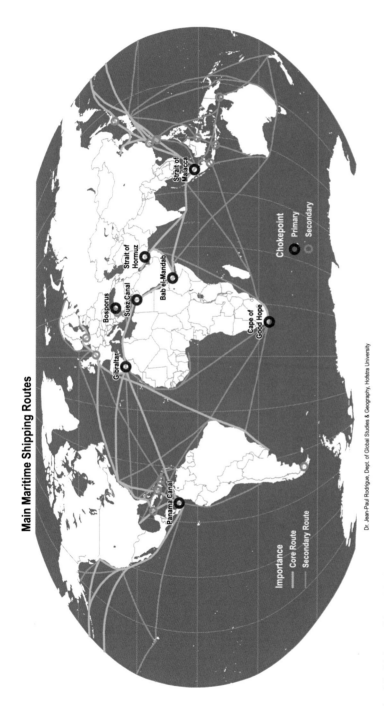

Fig. 3.1 Main maritime shipping routes, reprinted from [15], open access

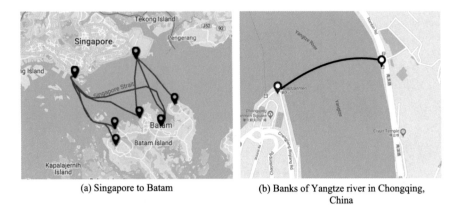

(a) Singapore to Batam   (b) Banks of Yangtze river in Chongqing,
                                         China

Fig. 3.2 Two cases of ferry routes

The practical cases include the first all-electric ferry "ampere" in Denmark, which is navigated only on batteries and can provide services with ZERO emission, and other cruise ships in Norled company.

3.2.1.2 Cruise Route

Cruise ships are mostly used for commercial purposes. Different from the short trips of ferries, cruise ships need to navigate for weeks with thousands of passengers and staff. Before the widespread of airlines, cruise ships were the only way for inter-continent traveling. Nowadays, cruise routes are mostly for tourism. Figure 3.3 gives

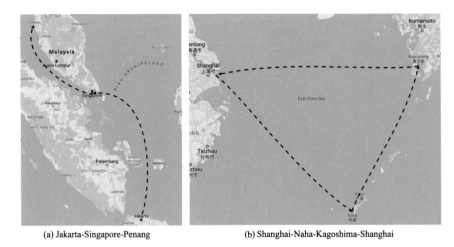

(a) Jakarta-Singapore-Penang   (b) Shanghai-Naha-Kagoshima-Shanghai

Fig. 3.3 Two cases of cruise routes

two examples of cruise routes.

Figure 3.3a shows a cruise route from Jakarta-Singapore-Penang, which is operated by the "Genting Dream" since 2016 [17]. The "Genting Dream" weights 151,300 dwt, and has 335 meters length, which can accommodate 4000 passengers and 2000 staff. The entire voyage lasts three days.

Figure 3.3b shows a cruise route from Shanghai-Naha-Kagoshima-Shanghai, which is operated by the "Norwegian Joy" since 2017 [18]. The "Norwegian Joy" weights 167,725 dwt, and has 333 meters length, which can accommodate 3800 passengers and 1800 staff. The entire voyage lasts six days.

In other places of the world, such as the Baltic Sea, the North Sea, the Caribbean Sea, and the Mediterranean Sea, there exist many types of cruise routes, and with the demand explosion of tourism, traveling by cruise ships will be more popular in the future.

3.2.1.3 Cargo/Container Ship Route

Nowadays, most of the oversea trading and a certain part of inland trading are based on maritime transportation, i.e., the cargo/container ships. Figure 3.4 shows a cargo/container ship route from Dalian, China to Aden, Yemen.

The total navigation time in Fig. 3.4 from Dalian to Aden takes 20 days. The oil tanker sails four times annually. Typical schedules are, the ship sets sail at 8:00 am on January 1st, April 1st, July 1st, and October 1st from Dalian, and returns on January 25th, April 25th, July 25th, and October 25th respectively from Aden [19].

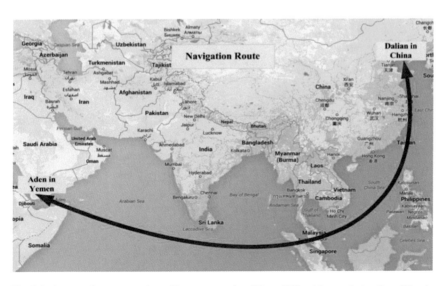

Fig. 3.4 A case of cargo-container ship route, reprinted from [19], with permission from Elsevier

3.2.2 Mathematical Model

Three main types of navigation routes have been described above. There exist many modeling methods for navigation tasks. In the following, a general complete model will be described in the first place, and a simplified model is then depicted in detail.

3.2.2.1 Time-Space Network Modeling

Generally, the decision variables during different navigation routes include: (1) the calls for the ports, i.e., choosing which port to berth in; (2) the navigation speed during each time-period; and (3) the total navigation time, i.e., determining the total navigation time to meet the requirements of customers. With the decision variables above, the mathematical model of navigation tasks can be shown by the following time-space network as Fig. 3.5, which is also shown in Refs. [20–22] as the navigation routing problems.

Assuming T is the total navigation time-period and is divided into $N_{|S|}$ time-intervals. Then the navigation task is modeled in a directed graph $G = (\bigcup_{(t=1)}^{(N_{|S|})} S_t, A_v)$, where S_t is the navigation point set (ports) which can be selected in the t-th time interval, and A_v is the arc set which connects two concessive time-intervals, i.e., t and $t + 1$. Each element is denoted as $a = (f, t) \in A_v = \left(s_t^f, s_{t+1}^t | \forall s_t^f \in S_t, \forall s_{t+1}^t \in S_{t+1}, t \in \left[1, N_{|S|} - 1\right]\right)$. For the t-th time-interval, the

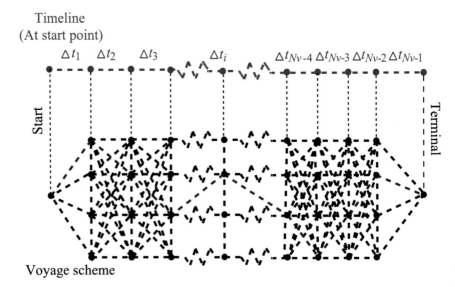

Fig. 3.5 Time-space network modeling of the navigation tasks

ship can choose an arc $a = \left(s_t^f, s_{t+1}^t \right)$ as the navigation route and $x_t^a = 1$ correspondingly, and in other cases, $x_t^a = 0$. The distance of (s_t^f, s_{t+1}^t) is denoted as l_a. The cruising speed is denoted as v_t^c in each navigation route. Then the navigation model can be shown as follows.

$$v_{t-1,t}^{min} \leq v_t^c \leq v_{t-1,t}^{min}, \ t \in [T_{t-1}, T_t] \tag{3.1}$$

$$T_t = T_{t-1} + T_t^{Na}, t \in N_{|S|} \tag{3.2}$$

$$T_t^{Na} = \Delta t_t = \frac{\sum x_t^a \cdot l_a}{v_t^c}, t \in N_{|S|} \tag{3.3}$$

$$T_t^{min} \leq T_t \leq T_t^{max}, t \in N_{|S|} \tag{3.4}$$

$$\sum_{s_t^f \in \delta^+(k)} x_t^a - \sum_{s_{t+1}^t \in \delta^-(k)} x_t^a = b_k, \forall k \in \bigcup_{t=1}^{N_{|S|}} S_t \tag{3.5}$$

where $v_{t-1,t}^{min}$, $v_{t-1,t}^{max}$ are the minimum and maximum navigation speed during time-period $t \in [T_{t-1}, T_t]$; T_t^{Na} is the consumed time of the navigation between $\left(s_{t-1}^f, s_t^t \right)$; T_t^{min}, T_t^{max} are the minimum and maximum consumed time when arriving at the port; $\delta^+(k)$ (resp. $\delta^-(k)$) denotes the set of arcs with the tail (resp. head) k; and $b_{s_t^f} = 1$, $b_{s_{t+1}^t} = -1$ and $b_k = 0$ for other cases.

Equation (3.1) represents the navigation speed should be within the upper and lower limits when navigation; Eq. (3.2) calculates the total consumed navigation time; Eq. (3.3) calculates the navigation time between two ports; Eq. (3.4) limits the total navigation time; Eq. (3.5) ensures the connectivity of the navigation scheme.

3.2.2.2 Simplified Modeling Method

In most cases, the navigation route is pre-determined and there is no need to re-schedule the route, and in those scenarios, the only action for determining an energy dispatch scheme is to adjust the navigation speed, and the above model in Sect. 3.2.2.1 can be simplified as Fig. 3.6. This simplification has been utilized in many research works [9, 10, 14, 23, 24].

Illustrated in Fig. 3.6, the voyage is divided into several time-intervals, and the duration of each time-interval is denoted as Δt. In each interval, the cruising speeds should be within the upper and lower bounds. Those time-intervals can be classified into two categories: (1) when the ship berths in the port (berthed intervals, denoted as T_b); (2) when the ship cruises within speed bounds (cruising intervals, denoted as T_c). During most time of T_c, the ship cruises around its nominal speed, while during

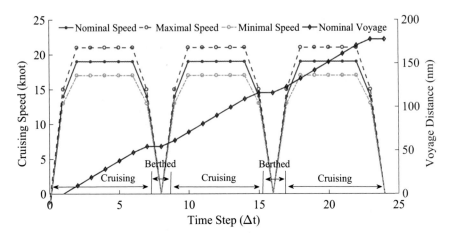

Fig. 3.6 Simplified navigation task model with a fixed route, reprinted from [9], with permission from IEEE

the time-intervals right approaching the port (partial-speed interval, denoted as T_p), it cruises at a slower speed. The relation between T_b, T_c is $T = T_b + T_c$, while the entire voyage horizon is $|\mathcal{K}|T$, and $|\mathcal{K}|$ is the number of ports.

As indicated in Fig. 3.6, the voyage distance between two consecutive time-intervals, i.e. t-th and t-1-th time-interval, is the accumulation of cruising speed with voyage duration Δt, which is represented as (3.6). Other constraints are shown in Eqs. (3.7)–(3.9).

$$Dist_t = Dist_{t-1} + v_t^c \cdot \Delta t \tag{3.6}$$

$$\left(1 - \delta_{D,k}^{max}\right) \cdot Dist_k^R \leq Dist_t \leq \left(1 + \delta_{D,k}^{max}\right) \cdot Dist_k^R, t \in T_p, t \neq |\mathcal{K}|T \tag{3.7}$$

$$Dist_{|\mathcal{K}|}^R \leq Dist_{|\mathcal{K}|T} \leq \left(1 + \delta_{D,|\mathcal{K}|}^{max}\right) \cdot Dist_{|\mathcal{K}|}^R \tag{3.8}$$

$$\begin{cases} \left(1 - \delta_v^{max}\right)v^n \leq v_t^c \leq \left(1 + \delta_v^{max}\right)v^n & \forall t \in T_c \\ \eta_p\left(1 - \delta_v^{max}\right)v^n \leq v_t^c \leq \eta_p\left(1 + \delta_v^{max}\right)v^n & \forall t \in T_p \\ v_t^c = 0 & \forall t \in T_b \end{cases} \tag{3.9}$$

where $Dist_t$ is the traveling distance at t-th time interval; $\delta_{D,k}^{max}$ is the maximum tolerance for traveling distance deviation; δ_v^{max} is the range of navigation speed; η_p is speed ratio when berthing out. This model is under the assumption of a fixed navigation route, which is suitable for energy dispatch analysis in many practical cases.

3.3 Energy Consumption

There are many energy sources in the maritime grids, such as diesel engine/generators (DGs), fuel cells (FCs), energy storage systems (ESSs), renewable energy generation, and so on. To achieve better environmental benefits, the minimization of energy consumption is an important management task.

3.3.1 Diesel Engines/Generators

DG acts as the main energy source for most of the commercial ships and the auxiliary energy sources of ports. DG can scale from several kilowatt to tens of megawatt in different application scenarios. Generally, DGs can be classified as three main types by their rotating speed: (1) slow-speed two-stroke DG; (2) medium-speed four-stroke DG; and (3) high-speed four-stroke DG. A general case of diesel engines in ship is shown as Fig. 3.7a [25].

The main differences between the above types of DGs are the rotating speed. The slow-speed two-stroke diesel engines are typically defined as the one with its rotating speed less than 400 rpm. The rotating speed of the medium-speed four-stroke diesel engines usually is limited within 400~1400 rpm, and the high-speed four-stroke diesel engine has more than 1400 rpm. In addition, the slow-speed two-stroke diesel engine only has two strokes in a full operation cycle, which leads to greater ability to export power than the other two types, meanwhile the size and capacity are also larger.

Currently, the efficiency of the slow-speed two-stroke diesel engine can achieve 52%, compared with 42% for common land-based vehicles. Typical specific fuel-oil consumption (SFOC) curves of different diesel engines are shown in Fig. 3.8.

(a) Diesel engine   (b) Direct propulsion

Fig. 3.7 Typical structure of diesel engines [25]

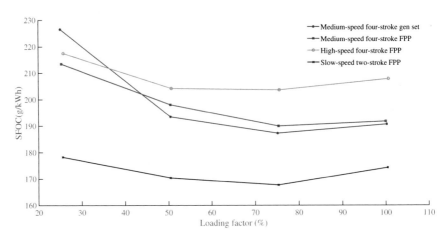

Fig. 3.8 Typical specific fuel-oil consumption (SFOC) curves

From Fig. 3.8, different diesel engines have their highest efficiency at 60~80% of rated power, and the energy consumptions under different power levels can be modeled as quadratic polynomial equations, shown as follows.

$$FC^{DG} = \hbar^2 \cdot (r_{DG})^2 + \hbar^1 \cdot r_{DG} + \hbar^0 \tag{3.10}$$

where FC^{DG} is the fuel consumption of the diesel engine; r_{DG} is the loading factor of the diesel engines, which is defined as $r_{DG} = P^t_{DG}/P^R_{DG}$, and P^t_{DG}, P^R_{DG} are the current power and rated power of the diesel engine; $\hbar^2$, $\hbar^1$, $\hbar^0$ are the coefficients, which can be derived from the experimental curves like Fig. 3.8.

In conventional cases, the slow-speed two-stroke diesel engines are mostly used as the primary energy sources in traditional ships for propulsion. The main reason for this wide usage is the ability to coordinate with various types of propellers. In traditional ships, the propulsion system is directly connected with the main diesel engine, shown as Fig. 3.7b. As we all known, the speed of propeller is low, only around 100 rpm. The low-speed diesel engines can therefore well accommodate various types of propellers. But for the medium and high-speed diesel engines, an extra speed reduction transmission system should be installed and brings 3~5% energy loss. However, in AESs, the propulsion system has no necessity to directly connect with the diesel engines, then the medium and high-speed diesel engines can be served as the main energy sources with convenience. In port-side applications, diesel engines usually act as the prime movers of power generators, and are the power backups for emergency usages, or sharing the power demand in peak hours.

3.3.2   Fuel Cell

Generally, fuel cell is a power source technology like diesel engines, but fuel cell directly transforms the chemical energy of fuel into electricity, thus in operating characteristics, the fuel cell is similar to the energy storage. For illustration, the fuel cell structure and its integration into ships are shown as Fig. 3.9a and b, respectively.

Since no spinning parts and no combustion process, fuel cells combine the characteristics of diesel engines and energy storages, i.e., the ability to continuously ouput power like diesel engines and the high efficiency like energy storages. In addition to the advantages of installment space and scalable capacity, fuel cells are viewed as a promising alternative energy source for the maritime grids, especially for the ships. In this field, fuel cell based on polymer exchange membrane (PEM) is the most mature technology, and has been already applied in ship applications [26, 27]. Additionally, fuel cells using conventional hydrocarbon fuels also have gained great concerns, such as Molten Carbonate Fuel Cell (MCFC) and Solid Oxide Fuel Cell (SOFC). Figure 3.10 gives the power characteristics of a methanol fuel cell [28].

From Fig. 3.10, we take the voltage curve when the fuel flow rate equals 12 mL/min as an example. The curve can be divided into phase I~III: (1) Phase I, electrochemical polarization zone; (2) Phase II, Ohm polarization zone; and (3) Phase III, concentration polarization zone. Phase I happens when the current is small, and in this phase, the electrochemical polarization effect enlarges the internal resistance of fuel cell. The voltage curve therefore has a deep drop. Then in phase II, the internal resistance is kept as a constant and the voltage characteristic follows Ohm's law. At last, in phase III, the concentration polarization effect dominates the process and further enlarges the internal resistance, meanwhile, the voltage suffers a deep drop. From the above Fig. 3.10, we can also find that the change of fuel flow rate has a significant effect on Phase III but smaller effects on Phase I and II, respectively.

In the power curves, the power of fuel cell will firstly increase with the current, then the power becomes saturated, at last in phase III, the power suffers a dramatic drop. For a well-designed fuel cell, the Maximum-Power-Point-Tracking Method

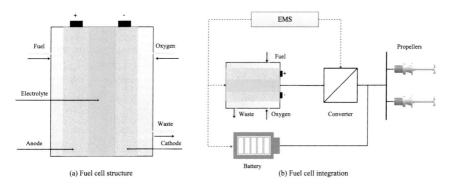

(a) Fuel cell structure   (b) Fuel cell integration

Fig. 3.9 Illustration of fuel cell

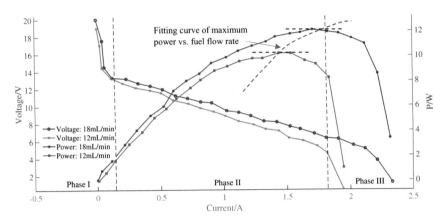

Fig. 3.10 Power characteristic curves of fuel cell [28]

(MPPT) will keep the fuel cell in the state of maximum power [29]. The fitting of the maximum power in each curve with the fuel flow rate, shown as the brown curve in Fig. 3.10, can represent the energy consumption model of a fuel cell. The formulation is shown as follows.

$$P^{FC} = g^2 \cdot (r_{FC})^2 + g^1 \cdot r_{FC} + g^0$$
$$FC^{FC} = p_f \cdot r_{FC} \cdot \Delta t \tag{3.11}$$

where P^{FC} is the power of fuel cell; r_{FC} is the fuel flow rate of fuel cell; g^2, g^1, g^0 are coefficients; FC^{FC} is the fuel consumption of fuel cell; p_f is the unit price of fuel; Δt is the length of time period.

3.3.3 Energy Storage

Generally, the operation of maritime grids includes the grid-connected and islanded modes, and most of maritime grids need to work in the shifting between two modes. For example, when the ship berths in a port and connects on cold-ironing equipment, the ship operates in grid-connected mode, and when the ship berths out, it works in islanded mode. For other ocean platforms, i.e., drilling platforms, offshore wind farms, they may work in different modes by cases. For example, when the offshore wind farms are connected with the main-land power system, they work in grid-connected mode and when they are connected with the islands, they work in islanded mode.

To keep the reliability in the above two modes, ESS is an important component for all types of maritime grids to act as an energy/power buffer between the generation-side and demand-side. In long-term timescale, an important application of ESS in maritime grids is to shave the peak load, which is shown in Fig. 3.11. When in peak

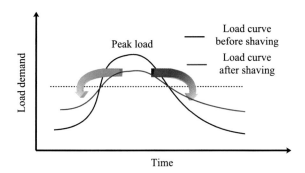

Fig. 3.11 Effects of energy storage in peak load shaving

load, ESS can discharge to share the power demand and in valley load, the ESS can charge and the load demand increases. Then the peak load can be viewed as "shaving" to other time periods and the maximum power demand can be reduced. This "load shaving" ability is very important to the maritime grids since most of the ships or ports don't have much power reserve, and the proper operation of ESS is essential to the reliability, security and stability of maritime grids.

The energy output of ESS is actually from other time-period, and the energy consumption of ESS is for the energy losses in the charging and discharging process, which is shown as follows.

$$
E_t^B = \begin{cases} E_{t-1}^B - P_{t-1}^{ESS} \cdot \eta_{ch} \cdot \Delta t \ \forall t \in \mathcal{T} \backslash 1, P_{t-1}^{ESS} < 0 \\ E_{t-1}^B - \frac{P_{t-1}^{ESS}}{\eta_{dis}} \cdot \Delta t \quad \forall t \in \mathcal{T} \backslash 1, P_{t-1}^{ESS} \geq 0 \end{cases} \tag{3.12}
$$

where E_t^B is the energy stored in the t-th time period; P_t^{ESS} is the power of ESS; Δt is the length of time-period; η_{ch}, η_{dis} are the charging/discharging efficiency of ESS.

3.3.4 Renewable Energy Generation

Renewable energy generation has been viewed as the solution to global fossil fuel depletion. Similar in maritime grids, renewable energy generation has also been gradually integrated. In Chap. 1, we have described many practical cases of renewable integration into ships. Here we give several cases to illustrate the development of renewable energy generation in ports.

The first case is from the Valencia port, Spain. This port plans to construct a breakwater dam and install tidal energy generation on it, shown as Fig. 3.12. The total capacity of the tidal energy generation can be 2.5 MW. A more detailed plan is proposed by the Houston port [13], which is shown in Fig. 3.13.

In Fig. 3.13, Spilman's island (area 6) is planned for the photovoltaic (PV) integration, and the PV power can be used to share the power demand of Houston port,

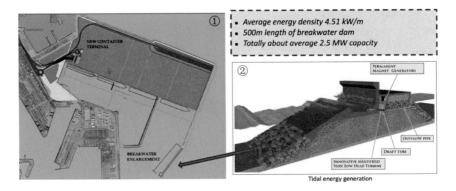

Fig. 3.12 Integration of tidal energy generation [30]

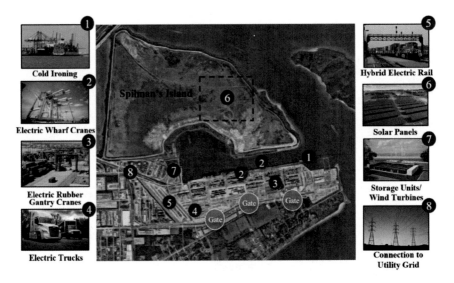

Fig. 3.13 Future development of Houston port, reprinted from [13], with permission from Elsevier

such as cold-ironing, various port cranes and electric transportation. In the future, renewable energy generation will play an even more significant role in maritime grids.

The renewable energy generation harvests different types of energy and transforms them into electricity. Its integration will reduce the usage of fossil fuel, which can be viewed as "negative fuel consumption". Its model can be shown as follow.

$$FC^{RE} = -E^{RE}/\eta_{FC} \tag{3.13}$$

where FC^{RE} is the fuel consumption reduction by renewable energy generation; E^{RE} is the total energy generated by the renewable energy generation; η_{FC} is the average efficiency of fossil fuel to electricity.

3.3.5 Main Grid

As above, the main grid is also an important energy source for maritime grids, especially for the ports. When the ship berths in a port and connects to the cold-ironing equipment, the main energy source also becomes the main grid. Generally, maritime grids will purchase electricity from the main grid according to the negotiated price, and the energy amount is measured at the substation. The model is shown as follows.

$$FC_{MG} = p_t \cdot E_{MG} \qquad (3.14)$$

where FC_{MG} is the price paid for electricity purchase; p_t is the electricity price in t-th time period; E_{MG} is the purchased electricity amount.

3.4 Gas Emission

As we all know, the great concern for gas emission in the maritime industry is the main motivation of maritime electrification, including the electrification of ships and ports. In the last decade, various energy regulations have progressively stimulated the innovations and targeted technology of all components that influencing the system performance from their design phases. As two main representatives of maritime grids, the management tasks of gas emission for the ships and ports are described as follows.

3.4.1 Gas Emission from Ships

3.4.1.1 Greenhouse Emission and Energy Efficiency Indexes

Currently, the plans of Energy Efficiency Design Index (EEDI) and the Energy Efficiency Operational Index (EEOI) are one part of the IMO's strategies to control the greenhouse emission from ships, which have two main roles: (1) providing a benchmark for comparing the energy efficiency of vessels; (2) setting a minimum required efficiency level for different ship types, size segments or cargo volumes.

The EEDI plan was first announced at the 62nd session of IMO's Marine Environment Protection Committee (MEPC 62) with the adoption of amendments to

MARPOL Annex VI, IMO [31]. After that, four important guidelines from IMO were enforced in the MEPC 63 in 2012 [32–35]. However, EEDI is used to measure greenhouse emission in the design phase for new ships. To implement this index for ships which have already been in service, EEOI is proposed as an amendment of EEDI in 2013 [36]. The general simplified formulas of EEDI and EEOI are shown as follows.

$$EEDI = \frac{(Engine\,power) \cdot SFC \cdot CF}{DWT \cdot speed} \quad (3.15)$$

$$EEOI = \frac{(Engine\,power) \cdot SFC \cdot CF}{(Cargo\,weight) \cdot speed} \quad (3.16)$$

where SFC is the specific fuel consumption of engine (g/kW); CF is the conversion factor of unit fuel to greenhouse emission; DWT is the deadweight of the ship; $speed$ is the navigation speed of ship. The difference between $EEDI$ and $EEOI$ is the deadweight to replace the cargo weight. Both the EEDI and EEOI can measure the greenhouse emission per unit transportation task.

From the above definitions, the ship which has higher energy efficiency will have lower values of EEOI and EEDI. A detailed description of EEDI and the meaning of each parts are shown as Fig. 3.14.

IMO also sets many reference lines for various ship types, and each type of ship should attain smaller EEDI than the reference line. The reference lines for some ship types are shown in Table 3.2.

IMO also sets many reduction targets in different time-period, i.e., (1) phase 0, 2013–2015; (2) phase 1, 2015–2020; (3) phase 2, 2020–2025; (4) phase 3, 2025 and later. Phase 1 requires a 10% reduction in the reference lines compared with phase 0, and phase 2 requires a 15~20% reduction in the reference lines compared with

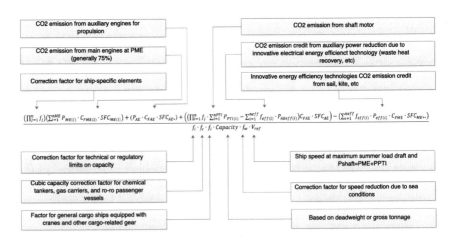

Fig. 3.14 The EEDI calculation formula

Table 3.2 Energy sources of different port-side equipment (data from [37])

Ship type		Reference line
Bulk carrier		$961.79 \times DWT^{-0.477}$
Gas carrier		$1120 \times DWT^{-0.456}$
Tanker		$1218.8 \times DWT^{-0.488}$
Container ship		$174.22 \times DWT^{-0.201}$
General cargo ship		$107.48 \times DWT^{-0.216}$
Refrigerated cargo carrier		$227.01 \times DWT^{-0.244}$
Combination carrier		$1219 \times DWT^{-0.488}$
Ro-Ro ship (Vehicle)	DWT/GT < 0.3	$(DWT/GT)^{-0.7} \times 780 \times DWT^{-0.471}$
	DWT/GT ≥ 0.3	$1812.63 \times DWT^{-0.471}$
Ro-Ro cargo ship		$1405.15 \times DWT^{-0.498}$
Ro-Ro passenger ship		$752.16 \times DWT^{-0.381}$
LNG carrier		$2253.7 \times DWT^{-0.474}$
Cruise passenger ship		$170.84 \times DWT^{-0.214}$

phase 0, and phase 3 requires a 30% reduction in the reference lines compared with phase 0. As we can see, the regulation for the greenhouse emission from ships will be even stricter in the future, and gradually becomes the primary management task for maritime grids.

3.4.1.2 NOx and SOx Emission from Ships

The concerns for both the NOx and SOx emission and some corresponding regulations are depicted in Sect. 1.4. Figure 3.15a and b respectively gives the typical emission characteristic of diesel engines for NOx and SOx [38].

From Fig. 3.15, the unit emission of both NOx and SOx will fall at first and then stabilize when the loading factor increased. In this sense, quadratic models can be formulated to represent the NOx and SOx emission from ships.

$$GE^{Ni} = g_2^{Ni} \cdot (r_{DG})^2 + g_1^{Ni} \cdot r_{DG} + g_0^{Ni} \tag{3.17}$$

$$GE^{Su} = g_2^{Su} \cdot (r_{DG})^2 + g_1^{Su} \cdot r_{DG} + g_0^{Su} \tag{3.18}$$

where GE^{Ni}, GE^{Su} are the NOx emission and SOx emission; g_2^{Ni}, g_1^{Ni}, g_0^{Ni} are the coefficients for NOx emission; g_2^{Su}, g_1^{Su}, g_0^{Su} are the coefficients for SOx emission; r_{DG} is the loading factor of diesel engine.

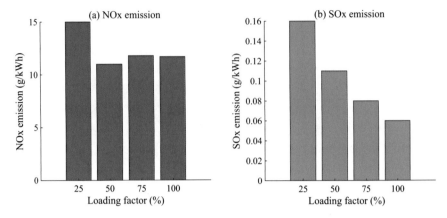

Fig. 3.15 NOx and SOx emission characteristics (data from [38])

3.4.2 Gas Emission from Ports

Generally, the gas emission from ports can be from three aspects, (1) maritime operation, including the approaching, hoteling and berthing-out of ships; (2) yard operation, including the operation of port-side logistic equipment, such as quay cranes, transferring vehicles and gantries; (3) generated hinterland logistic system operation, including the railways or land-based transportation system to transfer the cargo from the ports to the inland.

On the other side, the gas emission from ports has diversified types. Figures 3.16 and 3.17 respectively gives the breakdowns of different gas emissions in Taranto port [39] and Los Angel port [40].

From Fig. 3.16, CO_2 contributes to the majority of the total gas emission, and NOx, SOx, and particle mass (PM) are the other three main types of polluted gas emission. From Fig. 3.17, Ocean Going Vessels (OGV) are the main contributors for most of the gas emission, except the carbon monoxide (CO). Especially for the SOx emission, OGVs have contributed a 93.5% share. On the other hand, the cargo handling equipment is the highest contributor for CO emission, mainly for the incomplete combustion in the diesel engines of port cranes. At last, heavy-duty vehicles are also important CO_2 contributors, as well as a major NOx contributor.

To measure the gas emission from ports, CO_2, SOx and NOx are selected as the main representatives, and their calculations are similar and can be shown as Fig. 3.18.

Where GE is the total gas emission, i.e., CO_2, SOx, and NOx; $SFOC_{main}$, $SFOC_{aux}$, $SFOC_{eq}$ are the specific fuel oil consumptions (SFOCs) for the main engines, auxiliary engines, and the cargo handling equipment, respectively; EL_{main}, EL_{aux}, EL_{eq} are the average loading factors for the main engines, auxiliary engines, and the cargo handling equipment, respectively; EP_{main}, EP_{aux}, EP_{eq} are the capacities of the main engines, auxiliary engines, and the cargo handling equipment, respectively; D is the radius of the emission control area (ECA); V_s is the speed of

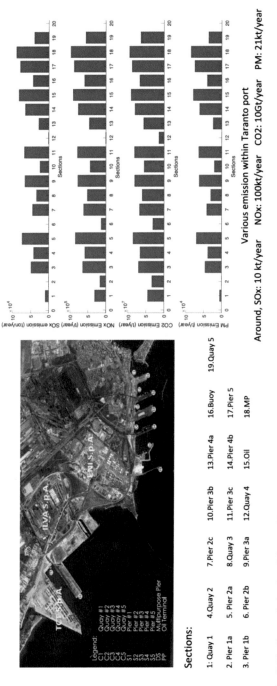

Fig. 3.16 Gas emission breakdown of Taranto port, Spain (data from [39])

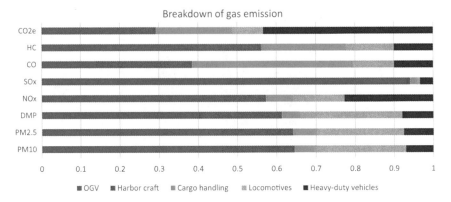

Fig. 3.17 Gas emission breakdown of Los Angeles port, US, reprinted from [40], open access

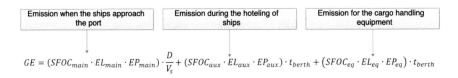

$$GE = (SFOC_{main} \cdot EL_{main} \cdot EP_{main}) \cdot \frac{D}{V_s} + (SFOC_{aux} \cdot EL_{aux} \cdot EP_{aux}) \cdot t_{berth} + \left(SFOC_{eq} \cdot EL_{eq} \cdot EP_{eq}\right) \cdot t_{berth}$$

Fig. 3.18 Calculation of gas emission from ports

ship when approaching the port. It should be noted that, Fig. 3.18 gives a general formula to calculate the gas emission of ports. The detailed model can be formulated after proposing the energy models of all the attached equipment.

3.5 Reliability Under Multiple Failures

During operation periods, maritime grids will face many types of failures, including equipment outages, short-circuit failures, and so on. In some severe scenarios, the failures may happen simultaneously and cause some serious consequences. To ensure the security of maritime grids, the reliability under multiple failures is an important management task. To simplify the modeling of multiple failures, only N-2 failures are considered in this book.

3.5.1 Multiple Failures in Ships

Different from the land-based maritime grids, such as ports, the ships are generally "islanded grids" when navigation. To ensure reliability, the ships are generally designed with two parallel buses. Some warships may even have four parallel buses

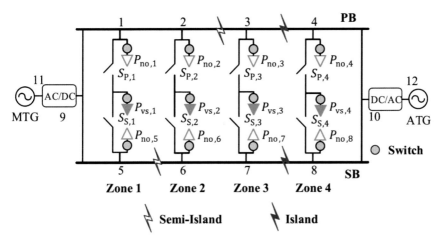

Fig. 3.19 Multiple failure types in ships, reprinted from [41], permission from IEEE

to increase the survivity. Figure 3.19 gives a typical topology with two parallel buses, i.e., PB and SB. Each load, i.e., $P_{no,1} \sim P_{no,8}$, $P_{vs,1} \sim P_{vs,4}$, can receive electricity from two buses, which means any one-bus failure will not cause any loss of load. Currently, the topology as Fig. 3.19 with two parallel buses is a common design for commercial ships.

The multiple failures in the ship power system can be classified into two types: (1) the semi-island mode. This mode has coupled zones between the island parts. For example, in Fig. 3.19, when in semi-island mode, the $P_{no,2}$ can still receive electricity from ATG by SB via the switch $S_{S,2}$, $S_{P,2}$. (2) island mode. This mode has no coupling zones at all, and the total system is divided into two islands, which is shown as the red failures in Fig. 3.19. In the above two severe multiple failure types, the island mode is more serious than the semi-island mode. Some of the loads have to be cut off if necessary.

3.5.2 Multiple Failures in Ports

The port grids are similar to conventional land-based distribution networks, which will supply various types of service to the berthed-in ships. A typical case is shown in Fig. 3.20, which has an electrical network, a water network, and a heat network. Three networks are coupled together since the water pump is driven by electricity and the combined heat power (CHP) generator is the source of the heat network.

As we can see, the networks in Fig. 3.20 are all radial ones, and any failure of equipment or branch will cause the loss of load demand (electricity, water, or heat). Then the network should be reconfigured to restore service. For example, when W3 is in failure, W4, W5 will have no water supply, then W2 and W5 should be reconfigured, then the water supply of W3~W5 can be restored.

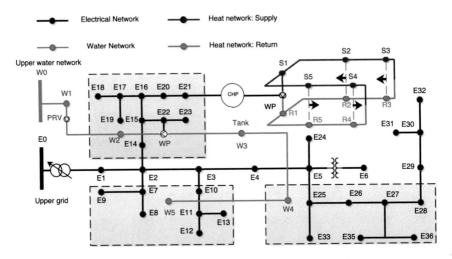

Fig. 3.20 A distribution network with multiple services (representing a port)

For multiple failures, there are two main types: (1) heterogeneous failures in different networks, such as one failure in electrical network and one failure in the water network; (2) homogeneous failures in the same network, such as two failures are both in the electrical network, or in water network. The latter failure mode has been well studied to improve network reliability [42, 43]. The former one involves different types of networks, which is not well studied at present.

3.5.3 Reliability Indexes

Reliability is the ability of the network to provide services in different operating conditions. Until now, there are many indexes to measure the reliability of different types of systems or networks. Table 3.3 gives some examples of reliability indexes,

Table 3.3 Conventional reliability indexes

Indexes	Explanations	Samples	Defintions
$P_{EPSF,i}$	The component failure probability in the i-th system	$P_{EPSF,i}^{(k)}$	$P_{EPSF,i} = \sum\limits_{k=1}^{k_{max}} \dfrac{P_{EPSF,i}^{(k)}}{k_{max}}$
$P_{EPLC,i}$	The probability of load shaving in the i-th system	$P_{EPLC,i}^{(k)}$	$P_{EPLC,i} = \sum\limits_{k=1}^{k_{max}} \dfrac{P_{EPLC,i}^{(k)}}{k_{max}}$
$P_{EDNS,i}$	The expected loss of load demand in the i-th system	$P_{EDNS,i}^{(k)}$	$P_{EDNS,i} = \sum\limits_{k=1}^{k_{max}} \dfrac{P_{EDNS,i}^{(k)}}{k_{max}}$

including the component failure probability, the probability of load shaving, and the expected loss of load demand.

The conventional calculation method for the reliability indexes is the Monte-Carlo simulation [44], which generate a set of scenarios and calculate the samples in each scenario. Then the reliability indexes are obtained by the average of samples. There are also plenty of analytical methods based on system approximation [45], which can calculate the reliability indexes more efficiently.

3.6 Lifecycle Cost

After the electrification of maritime grids, fuel cell and energy storage are both important equipment to improve the overall energy efficiency. Many research has investigated their applications and prove their benefits to the maritime grids. Due to the limits of current technologies, the fuel cell and energy storage cannot fully replace current power resources in the maritime grids. Therefore the fuel cell and energy storage need to operate coordinately with the other energy resources to supply the load demand. Additionally, the investment costs of fuel cell and energy storage are still high, and to reduce their overall operating cost, certain operating strategies should be implemented to extend their lifetime, and their lifetime model should be formulated in the first place.

3.6.1 Fuel Cell Lifetime Degradation Model

According to current research, there are many factors to influence the lifetime of fuel cell, including the operating temperature, humidity, and load profiles. Generally, fuel cells are installed in places with an advanced environmental control system, and the temperature and humidity can be sustained within a proper range [46]. Therefore the load profiles are the main factors that influence the lifetime of a fuel cell.

The load profiles which have influences on the fuel cell lifetime include (1) load changing; (2) start-up and shut-down; (3) idling; and (4) high load demand [47]. Then an empirical model for fuel cell lifetime can be shown as:

$$\Delta De_{FC} = K \cdot p((k_1 \cdot t_1 + k_2 \cdot n_1 + k_3 \cdot t_2) + \beta) \tag{3.19}$$

where ΔDe_{FC} is the degradation percentage of fuel cell; K is the degradation coefficient, which can be obtained by the degradation experiment; t_1, n_1, t_2 are the idle time, start-stop counts, and heavily loading time, respectively; β represents the natural decay rate. After defining the ΔDe_{FC}, the lifetime of fuel cell can be given by Eq. (3.20), and the operating cost of fuel cell can be given by Eq. (3.21).

$$L_{FC} = \frac{1 - EOL_{FC}}{\Delta De_{FC}} \tag{3.20}$$

$$Cost_{FC} = \frac{p_{FC} \cdot E_{FC}}{L_{FC}} \tag{3.21}$$

where L_{FC} is the lifetime of fuel cell; and EOL_{FC} is the end-of-life rate, generally 10%; $Cost_{FC}$ is the operating cost of fuel cell; p_{FC}, E_{FC} are the unit price and energy capacity of fuel cell, respectively.

3.6.2 Energy Storage Lifetime Degradation Model

Among all the energy storage technologies, battery is the most frequently used energy storage technologies for long-term energy management [48]. Furthermore, compared with other energy storage technologies, such as supercapacitors, flywheels, battery is more vulnerable and its lifetime is easier to be influenced by various operating conditions.

Similar to the fuel cell, there are also many factors to influence the lifetime of battery, such as temperature, humidity, and load profiles. Due to the installation of environmental control system, the load profiles are also the main factor on the battery lifetime. Among all the load profiles, the frequent discharging/charging events contribute to significant battery lifetime degradation, which is shown in Fig. 3.21.

In Fig. 3.21, the Depth of charge (DoD) is defined as d_b in a discharging or charging event, which is illustrated in Fig. 3.21a. The discharging or charging event is defined as the process between two concessive state-switching points (charging to discharging or vice versa), i.e., the continuous discharging or charging periods, in which the ESS maintains the single charging or discharging state and lasts for ΔT_{SC} with the average power $P_{Bat}^{avg, \Delta T_{sc}}$. During each charging or discharging event, d_b is defined as the difference between the SOCs before and after the event, which can be expressed as Eq. (3.22), where E^{Bat} is the energy capacity of the battery. The relation

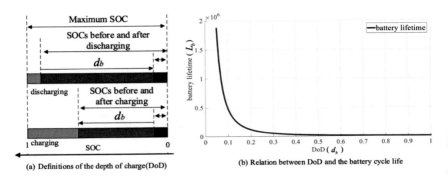

(a) Definitions of the depth of charge(DoD)   (b) Relation between DoD and the battery cycle life

Fig. 3.21 Depth of charge and the battery lifetime, reprinted from [9], permission from IEEE

between the DoD and the battery lifetime, denoted as L_b, is shown in Fig. 3.21b and the mathematical form in Eq. (3.23), where $a, b, c > 0$ are the fitting coefficients in Fig. 3.21b. At last, the operating cost of battery can be formulated as Eq. (3.24), where $Cost_{Bat}$ is the operating cost of battery, and p_{ES} is the unit investment of battery.

$$d_b(\Delta T_{SC}) = P^{avg}_{Bat} \cdot \Delta T_{SC} \big/ E^{Bat} \tag{3.22}$$

$$L_b(d_b) = a \cdot d_b^{-b} \cdot e^{-c \cdot d_b} \tag{3.23}$$

$$Cost_{Bat} = \frac{p_{ES} \cdot E^{Bat}}{L_b} \tag{3.24}$$

3.7 Quality of Service

Besides the economic benefits and allocated tasks, the quality of service (QoS) is also a vital management task for the maritime grids. There are many types of QoS, including the on-time rate of ships, the satisfactory level of passengers and ships. The on-time rate can be controlled by the management of navigation task in Sect. 3.2.1 and is not discussed here. The satisfactory levels of passengers and ships are described as below.

3.7.1 Comfort Level of Passengers

A cruise ship should provide heating load and hot water supply to the passengers. Equations (3.25) and (3.26) define the QoS of the above two services in a cruise ship. $\mathcal{T}$ is the total time period.

$$
\begin{aligned}
T_{Air,vio} &= T_{Air,vio.1} \cup T_{Air,vio.2} \\
T_{Air,vio.1} &= \left\{ T^{IN} \geq T^{IN,RE}_{max} \right\} \\
T_{Air,vio.2} &= \left\{ T^{IN} \leq T^{IN,RE}_{min} \right\} \\
T_{Wa,vio} &= \left\{ V^{HW} \leq (1 + \delta_{HW}) V^{SE} \right\}
\end{aligned} \tag{3.25}
$$

$$
\begin{aligned}
QoS_{Air} &= \frac{\left[\begin{array}{c} \int_{t \in T_{Air,vio.1}} \left(\left| T^{IN} - T^{IN,RE}_{max} \right| \right) \\ + \int_{t \in T_{Air,vio.2}} \left(\left| T^{IN} - T^{IN,RE}_{min} \right| \right) \end{array} \right]}{\int_{t \in T_{Air,vio}} \left(T^{IN,RE}_{max} - T^{IN,RE}_{min} \right)} \\
QoS_{Wa} &= \frac{\int_{t \in T_{wa,vio}} \left| V^{HW} - (1 + \delta_{HW}) V^{SE} \right|}{\int_{t \in T} \left| V^{SE} - \min(V^{SE}) \right|}
\end{aligned} \tag{3.26}
$$

where T^{IN} and V^{HW} represent the indoor temperature and hot water supply; $T_{max}^{IN,RE}$, $T_{min}^{IN,RE}$ are the maximal and minimal limits of the indoor temperature; V^{SE} is the required hot water demand; $T_{Air,vio}$ and $T_{Wa,vio}$ are defined as the time intervals or sub-intervals which violate the indoor temperature and hot water supply service requirement (tighter than the constraints). Equation (3.26) defines the QoS of indoor temperature and hot water supply, respectively. From the above definitions, the cruise ship with a lower QoS index will better satisfy the thermal load demand of the tourists. When the QoS index equals 0, the thermal load demand is met all the time.

3.7.2 Satisfaction Degree of Berthed-in Ships

For the berthed-in ships, the cold-ironing power and cargo handling are two main services provided by the port. Similar to the QoS of Eq. (3.27), the QoS for berthed-in ships can be defined as follow. $\mathcal{T}$ is the total time period.

$$
\begin{aligned}
T_{CP,vio} &= \left\{ P^{CP} \leq P_{min}^{CP} \right\} \\
QoS_{CP} &= \frac{\int_{t \in T_{CP,vio}} (|P^{CP} - P_{min}^{CP}|)}{\int_{t \in T_{CP,vio}} (|P_{min}^{CP}|)} \\
QoS_{CH} &= \frac{T^{CH} - T_{max}^{CH}}{T_{max}^{CH}}
\end{aligned}
\tag{3.27}
$$

where $T_{CP,vio}$ is the time intervals or sub-intervals which violates the cold-ironing power requirement; P^{CP} is the actual cold-ironing power; P_{min}^{CP} is the minimal required cold-ironing power; QoS_{CP} is the QoS of cold-ironing power; T^{CH} is the actual cargo handling time; T_{max}^{CH} is the maximal cargo handling time; QoS_{CH} is the QoS of cargo handling.

References

1. Dimopoulos, G., Stefanatos, C., Kakalis, P.: Exergy analysis and optimisation of a marine molten carbonate fuel cell system in simple and combined cycle configuration. Energy Convers. Manag. **107**, 10–21 (2016)
2. Trinklein, H., Parker, G., McCoy, T.: Modeling, optimization, and control of ship energy systems using exergy methods. Energy **191**, 116542 (2020)
3. Larsen, U., Nguyen, T., Knudsen, T., et al.: System analysis and optimisation of a Kalina split-cycle for waste heat recovery on large marine diesel engines. Energy **64**, 484–494 (2014)
4. Soato, M., Frangopoulos, A., Manente, G., et al.: Design optimization of ORC systems for waste heat recovery on board a LNG carrier. Energy Convers. Manag. **92**, 523–534 (2015)
5. Kyriakidis, F., Sørensen, K., Singh, S., et al.: Modeling and optimization of integrated exhaust gas recirculation and multi-stage waste heat recovery in marine engines. Energy Convers. Manag. **151**, 286–295 (2017)
6. Nour Eddine, A., Chalet, D., Faure, X., et al. Optimization and characterization of a thermoelectric generator prototype for marine engine application. Energy **143**, 682–695 (2018)

7. Yang, S., Chagas, B., Ordonez, C.: Modeling, cross-validation, and optimization of a shipboard integrated energy system cooling network. Appl. Therm. Eng. **145**, 516–527 (2018)

8. Chen, H., Zhang, Z., Guan, C., et al.: Optimization of sizing and frequency control in battery/supercapacitor hybrid energy storage system for fuel cell ship. Energy **197**, 117285 (2020)

9. Fang, S., Xu, Y., Li, Z., et al.: Two-step multi-objective optimization for hybrid energy storage system management in all-electric ships. IEEE Trans. Veh. Technol. **68**(4), 3361–3373 (2019)

10. Fang, S., Xu, Y., Li, Z., et al.: Optimal sizing of shipboard carbon capture system for maritime greenhouse emission control. IEEE Trans. Ind. Appl. **55**(6), 5543–5553 (2019)

11. Fang, S., Xu, Y., Wang, H., et al.: Robust operation of shipboard microgrids with multiple-battery energy storage system under navigation uncertainties. IEEE Trans. Veh. Technol. https://doi.org/10.1109/tvt.2020.3011117 (In press)

12. Shao, W., Zhou, P., Thong, K.: Development of a novel forward dynamic programming method for weather routing. J. Mar. Sci. Technol. **17**(2), 239–251 (2012)

13. Molavi, A., Shi, J., Wu, Y., et al.: Enabling smart ports through the integration of microgrids: a two-stage stochastic programming approach. Appl. Energy **258**, 114022 (2020)

14. Fang, S., Fang, Y., et al.: Optimal heterogeneous energy storage management for multi-energy cruise ships. IEEE Syst. J. (2020) (In press)

15. Rodrigue, J.-P., et al.: The Geography of Transport Systems, Hofstra University, Department of Global Studies & Geography, (2017). https://transportgeography.org

16. How to book ferries from Singapore to Batam. https://www.directferries.com/ferries_from_singapore_to_batam.htm

17. Genting Dream. https://en.wikipedia.org/wiki/Genting_Dream

18. Norwegian Joy. https://en.wikipedia.org/wiki/Norwegian_Joy

19. Lan, H., Wen, S., Hong, Y., et al.: Optimal sizing of hybrid PV/diesel/battery in ship power system. Appl. Energy **158**, 26–34 (2015)

20. Perera, L.P., Guedes Soares, C.: Weather routing and safe ship handling in the future of shipping. Ocean Eng. **130**, 684–695 (2017)

21. Krata, P., Szlapczynska, J.: Ship weather routing optimization with dynamic constraints based on reliable synchronous roll prediction. Ocean Eng. **150**, 124–137 (2018)

22. Walther, L., Rizvanolli, A., Wendebourg, M., et al.: Modeling and optimization algorithms in ship weather routing. Int. J. e-Navig. Marit. Econ. **4**, 31–45 (2016)

23. Fang, S., Gou, B., Wang, Y., et al.: Optimal hierarchical management of shipboard multi-battery energy storage system using a data-driven degradation model. IEEE Trans. Transp. Electrif. **5**(4), 1306–1318 (2019)

24. Fang, S., Xu, Y., et al.: Data-driven robust coordination of generation and demand-side in photovoltaic integrated all-electric ship microgrids. IEEE Trans. Power Syst. **35**(3), 1783–1795 (2019)

25. Important Things To Check In Ship's Engine Bedplate. https://www.marineinsight.com/tech/important-things-to-check-in-ships-engine-bedplate/

26. METHAPU Prototypes Methanol SOFC for Ships. Fuel Cells Bull. 5, 4–5. 2859(08)70190-1 (2008)

27. SFC Fuel Cells for US Army, Major Order from German Military. Fuel Cells Bull. 6, 4 (2012)

28. Mayur, M., Gerard, M., Schott, P., et al.: Lifetime prediction of a polymer electrolyte membrane fuel cell under automotive load cycling using a physically-based catalyst degradation model. Energies **11**, 2054 (2018)

29. Ahmadi, S., Abdi, S., Kakavand, M.: Maximum power point tracking of a proton exchange membrane fuel cell system using PSO-PID controller. Int. J. Hydrog. Energy **42**(32), 20430–20443 (2017)

30. Cascajo, R., García, E., et al.: Integration of marine wave energy converters into seaports: a case study in the port of valencia. Energies **12**, 787 (2019)

31. IMO: Resolution MEPC.203(62) Amendments to the Annex of the Protocol of 1997 to amend the International Convention for the Prevention of Pollution from Ships, 1973, as modified by the Protocol of 1978 (2011)

32. IMO: Resolution MEPC.212(63): 2012 Guidelines on the Method of Calculation of the Attained EEDI for new ships (2012)
33. IMO: Resolution MEPC.213(63), 2012 Guidelines for the development of a ship energy efficiency management plan (SEEMP) (2012)
34. IMO: Resolution MEPC.214(63): 2012 Guidelines on Survey and Certification of the EEDI, IMO MEPC (2012)
35. IMO: ResolutionMEPC.215(63): Guidelines for Calculation of Reference Lines for Use With the Energy Efficiency Design Index (EEDI) (2012)
36. IMO: Considerations of how to progress the matter of reduction of GHG emissions from ships. Note by the Secretariat. ISWG-GHG1/2. International Maritime Organization. London, 21 February (2017)
37. Class, N.K.: TEC-1048 survey and certification for EEDI and SEEMP required by the Amendments to ANNEX VI of MARPOL 73/7827 (2015)
38. Ergin, S., Durmaz, M., Kalender, S.: An experimental investigation on the effects of fuel additive on the exhaust emissions of a ferry. Proc. Inst. Mech. Eng. Part M J. Eng. Marit. Environ. 233(4), 1000–1006 (2019)
39. Adamo, F., Andriaal, G., et al.: Estimation of ship emissions in the port of Taranto. Measurement 47, 982–988 (2014)
40. Port of Los Angeles: Inventory of air emissions for calendar year 2016 (2017). https://www.portoflosangeles.org/environment/studies_reports.asp. Accessed Mar 2018
41. Xu, Q., Yang, B., Han, Q., et al.: Optimal power management for failure mode of MVDC microgrids in all-electric ships. IEEE Trans. Power Syst. 34(2), 1054–1067 (2019)
42. Duan, D., Ling, X., Wu, X., et al.: Reconfiguration of distribution network for loss reduction and reliability improvement based on an enhanced genetic algorithm. Int. J. Electr. Power Energy Syst. 64, 88–95 (2015)
43. Sotelo-Pichardo, C., Ponce-Ortega, J., Nápoles-Rivera, F., et al.: Optimal reconfiguration of water networks based on properties. Clean Technol. Environ. Policy 16(2), 303–328 (2014)
44. Billinton, R., Wang, P.: Teaching distribution system reliability evaluation using Monte Carlo simulation. IEEE Trans. Power Syst. 14(2), 397–403 (1999)
45. Billinton, R., Wang, P.: Distribution system reliability cost/worth analysis using analytical and sequential simulation techniques. IEEE Trans. Power Syst. 13(4), 1245–1250 (1998)
46. Hu, Z., Li, J., Xu, L., et al.: Multi-objective energy management optimization and parameter sizing for proton exchange membrane hybrid fuel cell vehicles. Energy Convers. Manag. 129, 108–121 (2016)
47. François, M., Yves, D., Sousso, K., et al.: Long-term assessment of economic plug-in hybrid electric vehicle battery lifetime degradation management through near optimal fuel cell load sharing. J. Power Sources 318, 270–282 (2016)
48. Díaz-González, F., Sumper, A., Gomis-Bellmunt, O., et al.: A review of energy storage technologies for wind power applications. Renew. Sustain. Energy Rev. 16(4), 2154–2171 (2012)

Chapter 4
Formulation and Solution of Maritime Grids Optimization

4.1 Synthesis-Design-Operation (SDO) Optimization

As a special energy system, the optimization of maritime grids can be considered as three levels similar to conventional land-based energy systems [1–3].

(1) Synthesis optimization. Synthesis is defined as the components used in the maritime grids and their connections. Via synthesis optimization, the optimal configuration of the maritime grids can be determined. For example, the ship hull design, electrical layout, and whether to integrate a component or not. Since the synthesis optimization answers the "Yes-or-No" questions and therefore involves certain binary decision variables.

(2) Design optimization or planning optimization. Design optimization is to determine the technical characteristics of components which are determined in synthesis optimization, such as the capacity and rated power. The difference between synthesis optimization and design optimization can be given by the well-known "siting and sizing" problems. The "siting problem" determines which type of components can be used and where to install them, which belongs to the synthesis optimization. Then the "sizing problem" determines the capacities of the installed components, which belongs to the design optimization. In power system research, design optimization is often named as planning optimization.

(3) Operation optimization. After the synthesis optimization and design optimization, the operation optimization determines the optimal operating states of each component under specified conditions. Taken the navigation speed as an example. The synthesis optimization determines the type of main engine and the design optimization determines the capacity of the main engine, then operation optimization determines the optimal loading levels to address different navigation scenarios, such as different wave and wind scenarios.

© The Author(s) 2021
S. Fang and H. Wang, *Optimization-Based Energy Management for Multi-energy Maritime Grids*, Springer Series on Naval Architecture, Marine Engineering, Shipbuilding and Shipping 11, https://doi.org/10.1007/978-981-33-6734-0_4

Three optimization problems are the basic problems for a maritime grid. They are abbreviated as Synthesis-Design-Operation (SDO) optimization [4]. It should be noted that only energy management optimization is within the scope of this book.

Many efforts have been devoted to this field to achieve the overall optimum of maritime grids. In this Chapter, the SDO optimization of maritime grids will be comprehensively reviewed in the following, (1) topologies of maritime grids; (2) typical SDO optimization problems; (3) compact form and solution methods. Compared with other review works, this Chapter firstly points out the significance of coordination between different maritime grids in SDO optimization.

4.2   Coordination Between Maritime Grids

Maritime grids are the offspring of extensive maritime electrification, and widely existing in ships, seaports, and various ocean platforms. Conventionally, maritime grids have very limited capacities and their optimizations also have limited influences on the overall system characteristics. For example, in conventional ships, the propulsion is directly driven by the main engines and the capacity of the corresponding ship power system is much smaller than the propulsion system. However, when a ship is fully electrified, the propulsion system becomes the electric load under the ship power system, and the energy management of ship power system can determine the economic and environmental characteristics of ship. Similarly, when a seaport is fully electrified, the energy management of seaport can coordinate both the logistic and electric systems to achieve better economic and environmental benefits. In this sense, with the development of maritime electrification, the energy management of maritime grids will play an even significant role in the future.

Generally, all the electric networks installed within harbor territory can be viewed as maritime grids, which act as the interface between ocean and land. To clarify the relationship between different maritime grids, we give Fig. 4.1 to show their operating framework. There are five types of maritime grids in Fig. 4.1, (1) wind farms, and (2) island microgrids, and (3) offshore platforms, and (4) seaport microgrids and (5) ship power systems. This illustration shows that the future maritime grids will be coupled with each other, and the coordinated optimization is necessary for future maritime grids.

(1) Offshore wind farms, can supply power to island microgrids, harbor city, offshore platforms, and seaport microgrid.
(2) Island microgrids, are islanded microgrids that are away from the main grid, which uses renewable energy and generators to supply the load.
(3) Seaport microgrid, is a grid-connected microgrid in an electrified seaport, which uses electricity to drive the port cranes and providing cold-ironing power to the berthed-in ships. Various renewable energy sources can be integrated into a seaport and excess electricity can sell to the main grid of harbor city.

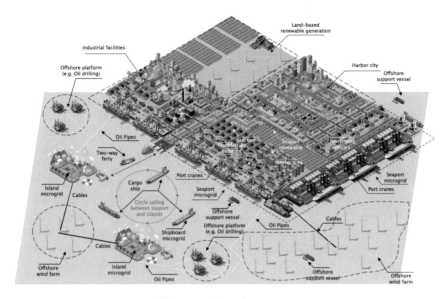

Fig. 4.1 Coordination between different maritime grids

(4) Offshore platforms, are islanded microgrids with many types of construction missions, such as fuel drilling, or underground cable construction. It should be noted that, offshore platforms may be connected with the islands and harbor city by underground pipes.

(5) Ship power systems. The ships can navigate between the islands, offshore platforms, and seaport to transfer fuel or other cargos. For example, the fuel produced by the offshore platform can be supplied to seaport by ships. It should be noted that, ships have different types, such the containers, cargo ships and the ferries for passenagers.

4.3 Topologies of Maritime Grids

Different types of maritime grids work in different conditions. For example, the notation of "CCO-HR(TEMP)+" in the American Bureau of Shipping (ABS) is for the ships which are working under low-temperature environment. "HR" is the emergency operating hours in the low-temperature environment (18 or 36 h). "TEMP" is the design service temperature, and "+" means that there is additional equipment for the crew for training in low-temperature [5]. The notation of "DPS" is for the dynamic position system of ships, which represents the ship has an automatic control system to maintain the position and heading at sea without external aid under specified conditions [5]. For the seaports, various equipment should be invested to serve the containers, cargo ships, or cruise ships, and so on, such as the port cranes to serve the containers, and the cold-ironing equipment for the berthed-in ships. The above designs are all determined by the synthesis optimization of ships.

With full electrification, maritime grids are multi-energy networks that use the electrical network as the backbone to supply various service networks, such as fuel flow network, thermal flow network, and water flow network [6]. In this sense, the synthesis optimization of maritime grids is mainly determining the topologies to achieve better performance. As two main representatives, the ship power system and the seaport microgrid are described in detail in this section.

4.3.1 Topologies of Ship Power Systems

For the ship power system, ABS has "R1", "R1-S", "R2", "R2-S" standards [5]. "R" is shorted for redundancy, "1" or "2" indicates the single/multiple sets of propellers and steering systems. "S" means the propulsion machines are located in separate compartments for emergency cases. Therefore, "R2-S" represents the ships that have multiple sets of propeller and steering system, and the propulsion machines are located in separate compartments. Among the above standards, "ABS-R2" is a conventional standard for the commercial ships under ABS, which means the ship power system can fully restore the serviceability when single failure. Figures 4.2, 4.3, 4.4, 4.5, 4.6 and 4.7 give some examples which follow the "ABS-R2" standard or above.

Ferries are small or medium-sized ships for passenager transportation, often using AC power supply with 690 V, which can carry hundreds to thousands of people, with a round-trip distance of tens of kilometers. For example, the world's first all-electric ferry, named as "Ampere", has been equipped with 2.6 MWh power batteries, reducing the use of 1 million liters diesel every year [7]. A typical illustration is shown in Fig. 4.2.

Cruise ships are large tourist ships that can carry thousands or even tens of thousands of people for several weeks, shown in Fig. 4.3. It usually uses the 11 kV AC power supply and is equipped with 4 or 6 generators. The cruise ship has many restaurants, playgrounds, cinemas, casinos, etc., which uses 440 V low-voltage power

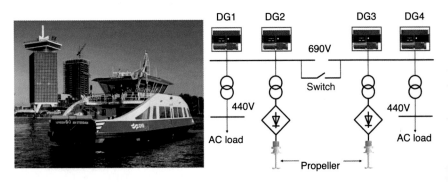

Fig. 4.2 All-electric ferry

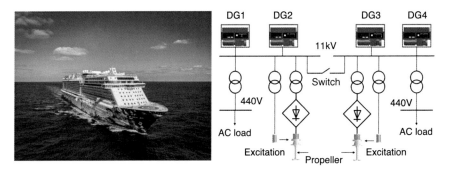

Fig. 4.3 All-electric cruise ship

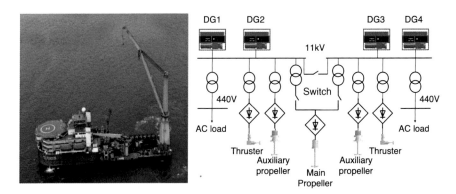

Fig. 4.4 All electric construction ship

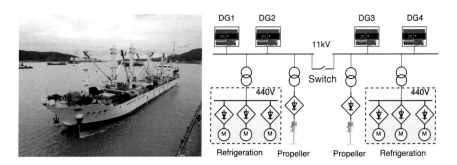

Fig. 4.5 All electric cold-chain transportation ship

supply. Due to the huge volume of cruise ships, the rated power of a single thruster can reach 20 MW. The total propulsion power of the Royal Caribbean's "Ocean Charm" is 97 MW [8].

Offshore construction ships are usually used for ocean-going operations, such as dredgers, oil-drilling, and fiber optic cable laying ships. Such ships require good

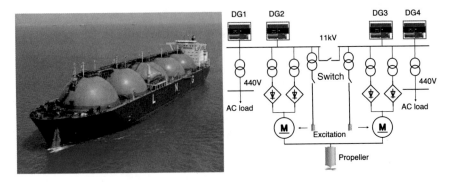

Fig. 4.6 All-electric LNG ship

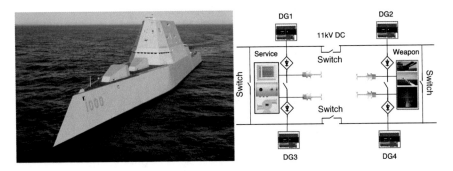

Fig. 4.7 All-electric warship

maneuverability, so they need propulsion systems with huge capacities, especially for the thrusters to meet ship steering and U-turn. A typical illustration is shown in Fig. 4.4.

Cold chain transportation ships usually transfer refrigerated containers and store all kinds of fresh food. This type of ship needs to provide a large amount of refrigeration load during navigation [9]. In Fig. 4.5, the refrigeration load is supplied by 440 V AC power.

LNG ships are mainly used to transport LNG. Unlike the cold chain transportation ships in Fig. 4.5, LNG ships do not directly supply refrigeration load, but mainly use the vaporization process of LNG to maintain Temperature ($-163\ °C$), and the vaporized natural gas is recovered through a reliquefaction device. This type of ship usually uses the 11 kV AC power supply [10]. Generally, this type of ships usually has a displacement of more than 100,000 tons, and the propellers need to be driven by several motors at the same time to ensure the maneuverability of the ship.

The last case is the warship which has multiple parallel buses (4 buses in Fig. 4.7) [9]. Four buses are backups to each other to ensure the survivability of warships on the battlefield.

In the future, with the development of all-electric ship, there will be more advanced topology designs for ship power systems, and the AC ship power system will be gradually replaced by the DC ship power system, which has larger capacities and more functionalities.

4.3.2 Topologies of Seaport Microgrids

Typically, the structure of seaport microgrid is similar to a land-based distribution network, which has (1) the main loop primary distribution network; (2) secondary loop-islands distribution network; and (3) tertiary distribution systems at specified voltage levels [11]. Figure 4.8 shows the structure of a seaport microgrid [12]. The main difference is the seaport has multiple redundant switches to ensure the power supply to critical loads, such as various port cranes and refrigeration.

With the electrification of seaport, seaport is required to provide more services to the berthed-in ships, such as the cold-ironing power. Additionally, the electrification of transferring vehicles is also another trend, which requires seaport to provide adequate charging facilities. Furthermore, harbor territory usually has much more plentiful renewable energy resources compared with inland, and the renewable generation integration into seaport to improve the energy efficiency is therefore a promising trend in the synthesis optimization of seaport microgrid. In this sense, the synthesis optimization of seaport microgrid can be viewed as an expansion planning problem. Ref. [13] analyzes the impact of cold-ironing power on the seaport and Ref. [14] analyzes the impact of renewable generation integration.

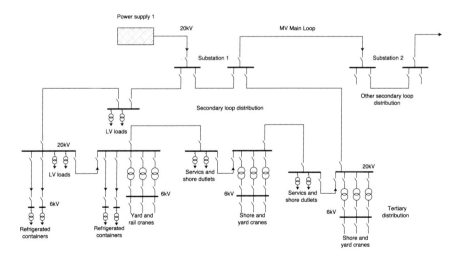

Fig. 4.8 Structure of a seaport microgrid

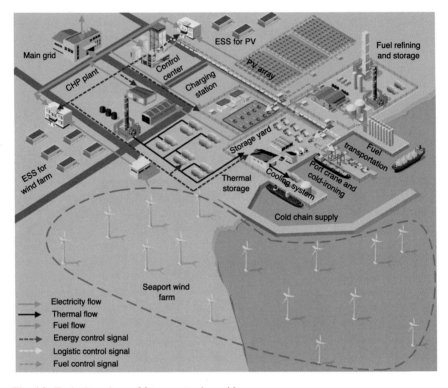

Fig. 4.9 Typical topology of future port microgrid

In the future, the seaport microgrid will become a multi-energy microgrid that involves electricity, thermal power, fossil fuel, and even water flow supply [6]. To clearly show the future operating framework, we re-draw Figs. 1.17 as 4.9. Different energy systems should be coordinately planned for an overall optimum.

4.3.3 Topologies of Other Maritime Grids

Besides the above two main representatives of maritime grids, i.e., ship power system and seaport microgrid, there still exist many other types of maritime grids, such as the island power system, drilling platform, or offshore oilfield. In this section, an island power system is shown in Fig. 4.10 as a representative [15].

At first, a power plant acts as the main power source of an island power system, and various renewable energy sources are integrated into this system, such as wind power and photovoltaic power. Additionally, several energy storage stations are used to improve the reliability.

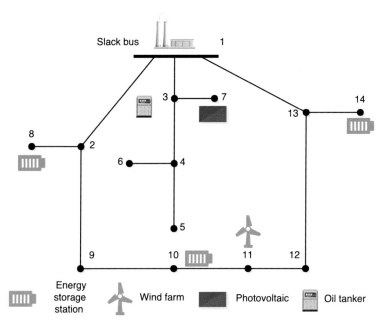

Fig. 4.10 An offshore island microgrid

4.4 Synthesis-Design-Operation Optimization of Maritime Grids

4.4.1 Synthesis Optimization for Maritime Grids

There are currently lots of research on the synthesis optimization of maritime grids. Here we give three cases to show their effects.

(1) Graph theory-based ship power system expansion

Nowadays, full electrification of ships is first implemented in Warships [16, 17] and may further expand to commercial applications [9]. As we know, ships may face various failures when navigation, such as malicious attacks on warships and component failures. To improve the resilience of a ship power system, Ref. [18] has proposed a graph theory-based ship power system expansion method to determine the optimal transmission line expansion strategy. The process is briefly described as follows.

Figure 4.11 gives a graph topology of an all-electric ship [19]. There are 22 buses and 29 lines in this graph. 4 generators are employed as the main power sources. 8 loads are classified as weapon load, propulsion load, radar load, control center, and hotel load according to the significance. The proposed model is to determine which lines should be installed for better resilience.

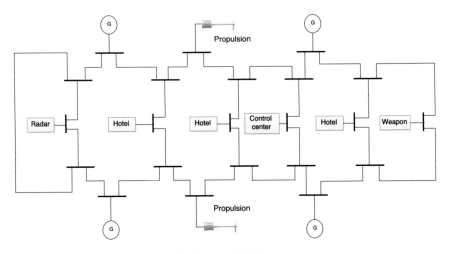

Fig. 4.11 The graph topology of an all-electric ship [19]

The proposed model has two objectives, (1) the weighted maximum flow from the generations to the loads, which is defined as (4.1), (2) Graph algebraic connectivity represented by the second smallest eigenvalue of Laplacian matrix.

$$\text{WSMF} = \sum_{ln} \omega_{ln} \cdot \sum_{gn} MF_{gn,ln} \tag{4.1}$$

where WSMF is the defined weighted maximum flow index; gn, ln are the index set for generations and loads; $MF_{gn,ln}$ is the maximum flow from the generations to the loads. The defined WSMF represents the maximum transmission capacity to the loads and can be acting as an important index to measure the resilience of ship power system.

In the case study, the proposed method is compared with the method of minimizing adding lines cost (MCR) [20]. The comparing results are shown in Table 4.1, and the simulation results bring two conclusions, (1) proper transmission line expansion can improve the resilience of ship power system; (2) the max-flow index is a useful index to measure the resilience of ship power system. From Table 4.1, the proposed model can reduce around 50% attacking scenarios which lead to load shedding.

Table 4.1 Load shedding results of different methods [18]

	1 Attacked line		2 Attacked lines	
	Attacking scenarios causing load shedding	Total attack scenarios	Attacking scenarios causing load shedding	Total attack scenarios
Original	10	27	233	351
MCR	6	29	167	406
Proposed	5	29	138	406

(2) Renewable generation expansion for Houston port

As the main interfaces between the ocean and inland, the environmental behaviors of seaports are always the concerns of the maritime industry [6]. With the electrification of seaports, massive renewable generation expansion in seaport has become reality. Ref. [14] proposes a model for the renewable generation planning and defines (1) smart energy index and (2) smart environmental index to measure the behaviors of seaports. The relevant parts with renewable energy integration are shown as follows.

$$SEgI_{RPG} = \frac{RS_{RPG} \cdot \sum P_{RPG} + RS_{MG} \cdot \sum (1 - \text{Pr}_{outage}) \cdot P_{MG}}{RS_T^{max}} \quad (4.2)$$

$$SEnI_{RPG} = -\frac{EM \cdot \sum P_{RPG}}{EM_T^{max}} \quad (4.3)$$

where $SEgI_{RPG}$ and $SEnI_{RPG}$ are the relevant parts of renewable power generation in smart energy index and smart environmental index; RS_{RPG}, RS_{MG} are the energy consumption ratios from renewable power generation and the main grid; P_{RPG} and P_{MG} are the power from renewable power generation and the main grid; Pr_{outage} is the outage percentage of the main grid; RS_T^{max} is the goal value of total renewable power generation within the seaport; EM is the average gas emission of unit power; EM_T^{max} is the goal value of total gas emission.

With the above two defined indexes, seaport can select a proper capacity of renewable power generation to achieve various economic and environmental management targets. The case study has shown that the gas mission of seaport can reduce more than 50% by the optimization of the proposed method, which can be a reference for future research.

(3) Structural optimization of an offshore oilfield power system

The ship power system and seaport microgrid are two main types of maritime grids, and there also exist various other maritime grids. The offshore oilfield power system is one representative that is studied by Ref. [15]. An offshore oilfield power system to be optimized is shown in Fig. 4.12a.

An offshore oilfield power system generally consists of an island and many drilling platforms. The island acting as the power source and a proper network structure should be planned to achieve (1) acceptable economic cost; (2) acceptable environmental behavior; and (3) acceptable reliability level. After solving the formulated model, the optimized structure is shown as Fig. 4.12b.

Practically, the drilling platforms may be away from an island. Therefore a more general case is the island is replaced by the mobile power plant. The mobile power system can move with the drilling platform when the mission is finished.

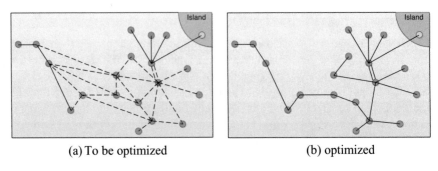

(a) To be optimized   (b) optimized

Fig. 4.12 Offshore oilfield power system. **a** To be optimized **b** optimized

4.4.2 Design and Operation Optimization for Maritime Grids

In this section, two cases are given to show the effects of design and operation optimization for maritime grids.

(1) Multi-agent energy management for a large port

Reference [21] proposes an energy management method based on a multi-agent system for a large electrified port. The agents in a port are shown in Fig. 4.13. The energy management process is simplified as follows.

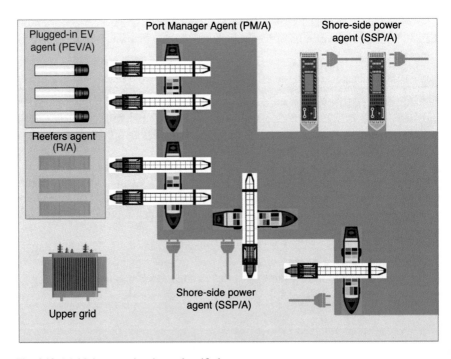

Fig. 4.13 Multiple agents in a large electrified port

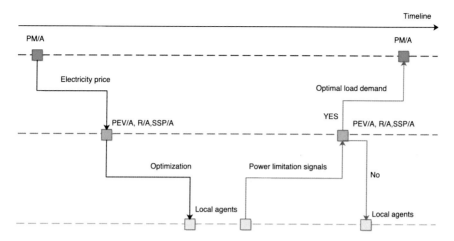

Fig. 4.14 Sequential energy management based on multiple agents

The overall port is under the control of the port manager agent (PM/A). PM/A aggregates the load demand within the port and communicates with the upper grid to determine the electricity price. The other agents include the plugged-in EV agent (PEV/A), which determines the charging/discharging of transferring vehicles, and the reefers agent (R/A), which determines the load demand of reefer containers, and shore-side power agent (SSP/A), which determines the cold-ironing power for each berthed-in ship. PEV/A, R/A, and SSP/A optimally dispatch the load demand of local agents (each component, such as one EV, or one reefer container) and then update with the PM/A. The overall process can be shown in the following Fig. 4.14.

PM/A sends electricity price to each agent (PEV/A, R/A, SSP/A), then each agent calculates its own optimal power demand plan and sends signals to each local agent of components. Each local agent determines if the load demand plan can be achieved. Then the "Yes/No" signals are sent back to PEV/A, R/A, and SSP/A. If all the local agents can achieve the optimal load demand, the total load demand under this agent will be sent to the PM/A. If not, the agent, i.e., PEV/A, R/A, SSP/A, will re-calculate the optimal load demand based on updated system conditions. This process will be repeated until convergence. This method has proved to be efficient and accurate in a real-world large electrified port. However, as an important part, the energy consumed by the port cranes are not considered in this research.

(2) Sizing of the shipboard gas capture system

We have mentioned the gas capture system in Chap. 1. Here we re-draw Fig. 1.11 to show the process of Ref. [22]. When the gas capture system is integrated into ships, the gas emission will be absorbed into storage and not emitted to the atmosphere. Before the wide usage of clean fuel, the gas capture technologies are viewed as feasible transition routes to control the gas emission.

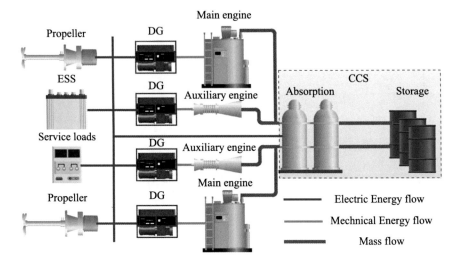

Fig. 4.15 Gas capture system into ships

Currently, the sulfur emission capture is the most mature technology among all the available technologies [23]. Lots of commercial applications have been implemented to meet the "2020 sulfur limit" [24]. The capture of carbon emission is a mature technology in land-based applications, but it still has many obstacles to be used in ships, such as the installment space, energy requirement, and so on. Other gas capture technologies, such as nitrogen capture and particle capture are all under investigation to find feasible implementations (Fig. 4.15).

The gas capture system integration will bring two problems, (1) what is the capacity of the gas capture system? and (2) what is the capacity of additional power sources to supply the gas capture system?

The first question is influenced by the environmental policies. For example, in 2020, IMO has launched the ever strictest sulfur limit policy, which requires to use 0.5% sulfur fuel. Then the installed gas capture system should have enough capacity to make the emitted gas has no more than 0.5% sulfur. In the future, the installed carbon capture system should also meet the global and regional carbon reduction goals. The second question is a design optimization for the ship power system. Since the original configuration of the ship does not have the gas capture system, the original generation system may not have enough capacity to supply the gas capture system. So extra power source, i.e., extra generator or energy storage, should be installed onboard. Ref. [22] formulates a sizing model to determine the above two questions. Its process is shown as the following Fig. 4.16.

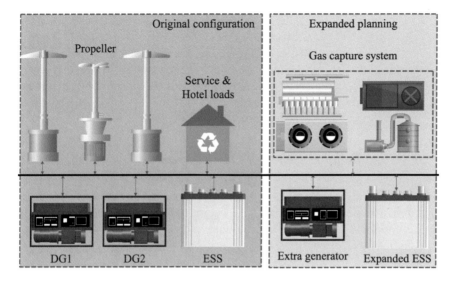

Fig. 4.16 Problems brought by gas capture system integration

4.5 Formulation and Solution of SDO Optimization

4.5.1 The Compact Form of SDO Optimization

In general terms, the compact form of SDO optimization for maritime grids can be shown as follow.

$$\min_{v,w,z} f(v, w, z) \tag{4.4}$$

where $f(v, w, z)$ is the management objective for SDO optimization, which is described in detail in Chap. 2; v is the set of decision variables for operation optimization, i.e., load factors of generators or engines, mass flow rates, pressure/temperature of streams, etc.; w is the set of decision variables for design optimization, i.e., nominal capacities of generators or engines, transmission limits of pipes or lines, etc.; z is the set of decision variables for synthesis optimization, which are generally binary variables to indicate the investment or non-investment of each component, i.e., with 1 value for investment and with 0 for non-investment.

For a complete SDO problem, Eq. (4.4) is under a set of constraints, including both equality and inequality constraints, to represent various limits in different scenarios.

$$h_i(v, w, z) = 0, i = 1, 2, 3 \ldots, I \tag{4.5}$$

$$g_j(v, w, z) \leq 0, j = 1, 2, 3 \ldots, J \tag{4.6}$$

A typical problem can involve one type, two types, or even three types of variables. For example, Refs. [1–3] involves three types of variable v, w, z, and Ref. [22] only involves two types. Generally, the SDO problems are non-linear and non-convex and very hard to be solved. Various methods have been proposed in this field to solve the SDO problems. In the following, the solution methods are classified into groups and then a decomposition-based method is described in detail for its usage in Chaps 4–7.

4.5.2 *Classification of the Solution Method*

The main classifications for solving the SDO problems are shown in Table 4.2 with some representative references, i.e., (1) mixed-integer linear programming, and (2) constrained non-linear programming, and (3) dynamic programming, and (4) evolutionary algorithm.

Table 4.2 Classifications for the solution methods of SDO problems

Models	Algorithms	References
Mixed-integer linear programming	Branch and bound	[25, 26]
Constrained non-linear programming	Generalized reduced gradient method	[27, 28]
	Sequential quadratic programming	[26, 29–31]
	Mixed-integer non-linear programming	[22, 32–36]
Dynamic programming	Sequential direct method	[3, 37]
	Deterministic dynamic programming	[38]
Evolutionary algorithm	Genetic algorithm	[39]
	Particle swarm optimization	[40, 41]
	Ant colony algorithm	[42]
	Whale optimization	[43]

4.5.3 Decomposition-Based Solution Method

In the following Chaps. 4–7, a decomposition-based solution method is proposed to solve a certain type of SDO problem, which is used in Refs. [22, 32–36] and belongs to the type of constrained non-linear programming. This type of SDO problem is shown in the following compact form.

$$\min_{v_1, v_2} f(v_1, v_2) \tag{4.7}$$

$$s.t. g_j^1(v_1) \leq 0, j = 1, 2, \ldots, J \tag{4.8}$$

$$h_1(v_1) = h_2(v_2) \tag{4.9}$$

$$g_i^2(v_2) \leq 0, i = 1, 2, \ldots, I \tag{4.10}$$

This problem belongs to the operation optimization, and v_1, v_2 are two types of operation variables, and usually belong to two different systems. For example in the navigation optimization of all-electric ships, v_1 represents the energy-related variables and v_2 related to the speed variables.

In the above formulation, Eqs. (4.8), (4.10) are the constraints in two different systems. For example in the navigation optimization of all-electric ships, Eq. (4.8) is the constraint set for energy and Eq. (4.10) is the constraint set for speed, and they are related by Eq. (4.9). This is a special type of maritime grid optimization problems since the couple between two systems only lies on Eq. (4.9).

In this book, this type of problem can be solved by a decomposition-based method and divided the original model into two levels as Eqs. (4.11) and (4.12).

$$\begin{gathered} \min_{v_1} f\left(v_1, v_2^*\right) \\ s.t. g_j^1(v_1) \leq 0, j = 1, 2, \ldots, J \\ h_1(v_1) = h_2\left(v_2^*\right) \end{gathered} \tag{4.11}$$

$$\begin{gathered} \min_{v_2} f_{aux}\left(v_1^*, v_2\right) \\ s.t. g_i^2(v_2) \leq 0, i = 1, 2, \ldots, I \end{gathered} \tag{4.12}$$

Equation (4.11) refers to the upper level, and in this level, decision variables v_2 are viewed as constant variables, which are updated in the lower level, i.e., Eq. (4.12). In Eq. (4.12), f_{aux} is an auxiliary objective function, which represents a management target, such as minimization of voyage deviation, or minimization of voyage period, and so on. With the above decomposition, the original problem is decomposed into two simplified sub-problems.

In literature, Refs. [22, 32–36] solve the energy management problem for all-electric ships. Equation (4.8) is the energy-related constraints and Eq. (4.10) is the speed-related constraints, and Eq. (4.9) is the speed-energy relationship, i.e., a cubic polynomial constraint. With the above decomposition, a non-linear and non-convex original problem is reformulated as a quadratic and linear programming problems, respectively, and therefore can be solved efficiently. In the following Chapters, this method will be used to solve various practical problems. Ref. [44] give a general method to select the parameters of this solution method.

References

1. Frangopoulos, C.: Optimization of synthesis-design-operation of a cogeneration system by the Intelligent Functional Approach. Int. J. Energy Environ. Econ. **1**, 275–287 (1991)
2. Sakalis, G., Frangopoulos, C.: Inter-temporal optimization of synthesis, design and operation of integrated energy systems of ships: general method and application on a system with Diesel main engine. Appl. Energy **226**, 991–1008 (2018)
3. Sakalis, G., Tzortzis, G., Frangopoulos, C.: Inter-temporal static and dynamic optimization of synthesis, design and operation of integrated energy systems of ships. Energies **12**, 893 (2019)
4. Frangopoulos, C.: Developments, trends, and challenges in optimization of ship energy systems. Appl. Sci. **10**, 4639 (2020)
5. ABS Notations and Symbols–20 February 2020. https://ww2.eagle.org/en/rules-and-resources/rules-and-guides.html
6. Fang, S., Wang, Y., et al.: Toward future green maritime transportation: an overview of seaport microgrids and all-electric ships. IEEE Trans. Veh. Technol. **69**(1), 207–220 (2020)
7. Norway electric ferry cuts emissions by 95%, costs by 80%. https://reneweconomy.com.au/norway-electric-ferry-cuts-emissions-95-costs-80-65811/
8. Hansen, J.F., Wendt, F.: History and state of the art in commercial electric ship propulsion, integrated power systems, and future trends. Proc. IEEE **103**(12), 2229–2242 (2015)
9. Kumar, D., Zare, F.: A comprehensive review of maritime microgrids: system architectures, energy efficiency, power quality, and regulations. IEEE Access **7**, 67249–67277 (2019)
10. Fernández, I., Manuel, R., Javier, R., et al.: Review of propulsion systems on LNG carriers. Renew. Sustain. Energy Rev. **67**, 1395–1411 (2017)
11. Parise, G., Hesla, E., Rifaat, R.: Architecture impact on integrity of electrical installations: cut&tie rule, ring configuration, floating node. IEEE Trans. Ind. Appl. **45**(5), 1903–1909 (2009)
12. Parise, G., Parise, L., Martirano, L., et al.: Wise port and business energy management: port facilities, electrical power distribution. IEEE Trans. Ind. Appl. **52**(1), 18–24 (2015)
13. Thalis, Z., Jacob, N., et al.: Evaluation of cold ironing and speed reduction policies to reduce ship emissions near and at ports. Marit. Econ. Logist. **16**(4), 371–398 (2014)
14. Molavia, A., Shib, J., et al.: Enabling smart ports through the integration of microgrids: a two-stage stochastic programming approach. Appl. Energy **258**, 114022 (2020)
15. Long, G., Yu, Q., Li, X., et al.: Structural optimization of offshore oilfield interconnected power system considering reliability. In press, Power System Technology (2020)
16. Doerry, N.: Zonal ship design. Naval Eng. J. **118**(1), 39–53 (2006)
17. DDG 1000 class destroyer DDG 1000 program highlights. In Sea Air Space-The Navy League's Global Maritime Exposition, pp. 1–8. National Harbor, Maryland, United States (Apr. 2013)
18. Lai, K., Illindala, M.: Graph theory based shipboard power system expansion strategy for enhanced resilience. IEEE Trans. Ind. Appl. **54**(6), 5691–5699 (2018)

19. Jin, Z., Sulligoi, G., Cuzner, R., et al.: Next-generation shipboard dc power system: introduction smart grid and dc microgrid technologies into maritime electrical networks. IEEE Electrif. Mag. **4**(2), 45–57 (2016)
20. Zhang, X., Tomsovic, K., Dimitrovski, A.: Security constrained multistage transmission expansion planning considering a continuously variable series reactor. IEEE Trans. Power Syst. **32**(6), 4442–4450 (2017)
21. Kanellos, F., Volanis, E., Hatziargyriou, N.: Power management method for large ports with multi-agent systems. IEEE Trans. Smart Grid **10**(2), 1259–1268 (2017)
22. Fang, S., Xu, Y., et al.: Optimal sizing of shipboard carbon capture system for maritime greenhouse emission control. IEEE Trans. Ind. Appl. **55**(6), 5543–5553 (2019)
23. Ryu, Y., Kim, H., Cho, G., et al.: A study on the installation of SCR system for generator diesel engine of existing ship. J. Korean Soc. Mar. Eng. **39**(4), 412–417 (2015)
24. Fan, L., Gu, B.: Impacts of the increasingly strict sulfur limit on compliance option choices: the case study of chinese SECA. Sustainability **12**(1), 165 (2020)
25. Solem, S., Fagerholt, K., Erikstad, S., et al.: Optimization of diesel electric machinery system configuration in conceptual ship design. J. Marit. Sci. Technol. **20**, 406–416 (2015)
26. Baldi, F., Ahlgren, F., Melino, F., et al.: Optimal load allocation of complex ship power plants. Energy Convers. Manag. **124**, 344–356 (2016)
27. Kalikatzarakis, M., Frangopoulos, C.: Multi-criteria selection and thermo-economic optimization of organic Rankine cycle system for a marine application. Int. J. Thermodyn. **18**, 133–141 (2015)
28. Kalikatzarakis, M., Frangopoulos, C.: Thermo-economic optimization of synthesis, design and operation of a marine organic Rankine cycle system. Proc. Inst. Mech. Eng. Part M J. Eng. Marit. Environ. **231**, 137–152 (2017)
29. Dimopoulos, G., Stefanatos, I., Kakalis, N.: Exergy analysis and optimisation of a steam methane pre-reforming system. Energy **58**, 17–27 (2013)
30. Soato, M., Frangopoulos, C., Manente, G., et al.: Design optimization of ORC systems for waste heat recovery on board a LNG carrier. Energy Convers. Manag. **92**, 523–534 (2015)
31. Dimopoulos, G., Stefanatos, I., Kakalis, N.: Exergy analysis and optimisation of a marine molten carbonate fuel cell system in simple and combined cycle configuration. Energy Convers. Manag. **107**, 10–21 (2016)
32. Fang, S., Xu, Y., Li, Z., et al.: Two-step multi-objective optimization for hybrid energy storage system management in all-electric ships. IEEE Trans. Veh. Technol. **68**(4), 3361–3373 (2019)
33. Fang, S., Xu, Y., Wang, H., et al.: Robust operation of shipboard microgrids with multiple-battery energy storage system under navigation uncertainties. IEEE Trans. Veh. Technol. https://doi.org/10.1109/tvt.2020.3011117. (In press)
34. Fang, S., Fang, Y., et al.: Optimal heterogeneous energy storage management for multi-energy cruise ships. IEEE Syst. J. 2020. (In press)
35. Fang, S., Gou, B., Wang, Y., et al.: Optimal hierarchical management of shipboard multi-battery energy storage system using a data-driven degradation model. IEEE Trans. Transp. Electrif. **5**(4), 1306–1318 (2019)
36. Fang, S., Xu, Y., et al.: Data-driven robust coordination of generation and demand-side in photovoltaic integrated all-electric ship microgrids. IEEE Trans. Power Syst. **35**(3), 1783–1795 (2019)
37. Tzortzis, G., Frangopoulos, C.: Dynamic optimization of synthesis, design and operation of marine energy systems. Proc. Inst. Mech. Eng. Part M J. Eng. Marit. Environ. **233**, 454–473 (2019)
38. Wu, P., Bucknall, R.: Hybrid fuel cell and battery propulsion system modelling and multi-objective optimization for a coastal ferry. Int. J. Hydrog. Energy **45**, 3193–3208 (2020)

39. Shang, C., Srinivasan, D., Reindl, T.: Economic and environmental generation and voyage scheduling of all-electric ships. IEEE Trans. Power Syst. **31**(5), 4087–4096 (2016)
40. Kim, D., Hwang, C., Gundersen, T., et al.: Process design and economic optimization of boil gas reliquefaction systems for LNG carriers. Energy **173**, 1119–1129 (2019)
41. Koo, J., Oh, S., Choi, Y., et al.: Optimization of an organic rankine cycle system for an LNG-powered ship. Energies **2019**, 12 (1933)
42. Yang, S., Chagas, M., Ordonez, J.: Modeling, cross-validation, and optimization of a shipboard integrated energy system cooling network. Appl. Therm. Eng. **145**, 516–527 (2018)
43. Chen, H., Zhang, Z., Guan, C., et al.: Optimization of sizing and frequency control in battery/supercapacitor hybrid energy storage system for fuel cell ship. Energy **197**, 117285 (2020)
44. Fang, S., Xu, Y.: Multi-objective coordinated scheduling of energy and flight for hybrid electric unmanned aircraft microgrids. IEEE Trans. Ind. Electron. **66**(7), 5686–5695 (2019)

Chapter 5
Energy Management of Maritime Grids Under Uncertainties

5.1 Introductions of Uncertainties in Maritime Grids

5.1.1 Different Types of Uncertainties

There are many types of uncertainties during the operation of maritime grids, i.e., demand-side uncertainties, generation-side uncertainties, and failure uncertainties, which are shown in Fig. 5.1.

Generally, navigation uncertainties are the main sources of demand-side uncertainties, such as the uncertain wave and wind and the adverse weather conditions. As we have illustrated in former Chap. 2, there are different management tasks of maritime grids, and the navigation uncertainties therefore can bring uncertainties to the demand, such as the propulsion load in ships and the corresponding calls-for-service delay for berthing.

For the propulsion load, conventional uncertain wave and wind will add navigation resistance and cause speed loss. To ensure the on-time rates, the power generation system requires a certain power reserve, noted as "sea margin" [1]. Table 5.1 shows the "sea margins" in the main navigation route around this world.

From the above table, the "sea margins" are generally within the range of "20%–30%", which represents a general ship design should at least have 30% power reserve [2]. This power reserve range has provided the flexibility for the maritime grids to accommodate navigation uncertainties towards economic and environmental objectives, and also gives the necessity of optimal energy management. When the navigation uncertainties continuously increasing, the route may become not suitable for navigation, and this type of navigation uncertainties is the "adverse weather conditions", the ships need to change another route for safety, which refers to the "weather routing" problems [3–5]. Additionally, navigation uncertainties will bring calls-for-service delays, which means the ships cannot arrive at the mission point at the scheduling time, and the service will be delayed. For example, the pre-scheduled

© The Author(s) 2021
S. Fang and H. Wang, *Optimization-Based Energy Management for Multi-energy Maritime Grids*, Springer Series on Naval Architecture, Marine Engineering, Shipbuilding and Shipping 11, https://doi.org/10.1007/978-981-33-6734-0_5

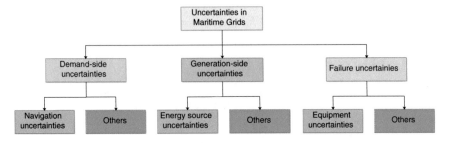

Fig. 5.1 Classification of uncertainties in maritime grids

Table 5.1 "Sea margins" for the main navigation route [2]	Navigation route	Sea margin
	North Atlantic, heading west	25%–35%
	North Atlantic, heading east	20%–25%
	Europe-Australia	20%–25%
	Europe-East Asia	20%–25%
	Pacific	20%–30%

berth position for a delayed ship may stay idle state till the ship arrives, and the electric and logistic service will also be postponed, which brings uncertainties to the operation of the seaport.

The energy source uncertainties are the main sources of generation-side uncertainties. In conventional operating scenarios, the uncertainties of energy sources are quite limited since the main energy sources, such as diesel engines or gas turbines are highly controllable. However, in recent years, various types of renewable energy are integrated into maritime grids, and the inherent intermittency brings lots of operating uncertainties to the maritime grids, such as the photovoltaic energy in ships, and the offshore wind farms for island microgrid. Those types of uncertainties should be addressed to reduce the operating risks of maritime grids. Another type of energy source uncertainties is the main grid uncertainties from the uncertain electricity price and the main grid failures.

The equipment uncertainties generally include two types, the first one is for the failure and the second one is for the scheduled maintenance or replacement. Their difference is the failure occurs unexpectedly and the system needs to act for correction, and the scheduled uncertainties give much longer time for the system to re-schedule the operating plan.

The main classifications of uncertainties in maritime grids are illustrated above, and the uncertainties bring enormous operating risk to the maritime grids, and proper operating strategies should be promoted to mitigate their influence.

5.1.2 Effects of Electrification for Uncertainties

The above uncertainties have perplexed the maritime industry for a long time. For example, the adverse weather conditions have been viewed as one of the main reason for ocean accidents, for example, the accident of *Svendborg Maersk* in 2014 [6]. The equipment failures are also viewed as the enemy to the system reliability [7, 8]. Lots of control strategies have been studied to prevent their harmful effects, such as spare parts optimization [9] and system reconfiguration [8]. With the extensive electrification, the maritime grids become a highly coupled multi-energy system, and all the system resources can be used to mitigate one type of uncertainty. Here we give two examples to show the effects of electrification for uncertainties.

In conventional ships, the propulsion system is directly driven by the main engine by a gearbox, and the other onboard components are supplied by the shipboard power system, shown as Fig. 5.2.

The main engines cannot freely adjust their outputs, and only several gear positions can be selected, such as 1/2 full power or 1/3 full power. As a result, this type of adjustment is coarse and lacks flexibility for conventional navigation uncertainties, since, for most cases, the speed loss by navigation uncertainties is only 10%~15% of the total speed [2]. With extensive electrification, the propulsion system can be quickly responding to the navigation uncertainties due to the superior rotation regulation performance of electric machines [10, 11]. In this sense, the electrification can make ships navigate in a more steady speed range.

For conventional seaports, the logistic equipment consumes most of the energy consumption and they may be not driven by electricity, such as the rubber-tire gantry may be driven by diesel engines [12]. In some cases, the rest system may not consume all renewable energy integration. For example, the Jurong port of Singapore has 9.5 MW photovoltaic energy integration in 2016 [13], but some of them may be wasted in some time periods. However, with fully electrified, the maritime grids will have a larger capacity to accommodate the energy source uncertainties. Furthermore,

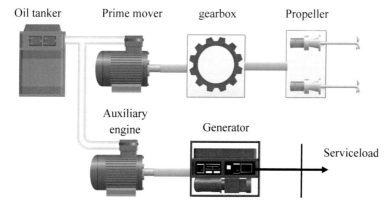

Fig. 5.2 Topology of conventional ships

the auxiliary equipment brought by electrification, such as batteries, combined heat-cooling power generators, can further enhance the reliability under main grid failures.

In summary, with full electrification, the maritime grids will have larger capacity and more resources to withstand different types of uncertainties, and with proper energy management, the economy, environment, and reliability objectives can be better achieved.

5.2 Navigation Uncertainties

5.2.1 Uncertain Wave and Wind

Generally in calm water, the propulsion load of ship has a cubic relationship with the cruising speed, shown as Fig. 5.3. The propulsion load will gradually increase with the speed and finally hits the "wave wall", and the maximum cruising speed achieves. When a ship sails on the sea, the wave and wind will add extra resistance and bring speed losses [2], and the wave wall will be moved to the left.

From Fig. 5.3, when considering the wave and wind, the cruising speed under the same propulsion load will decrease, and this refers to the "speed loss". To mitigate this speed loss, the main engine of ships should have adequate "sea margin", usually more than 15% by different navigation routes and seasons, as shown in Table 5.1. For example, in the route between Japan-Canada, the added resistance may scale up to 220% in some seasons, and the average is about 100% [14], and for most cases, the resistance in summer increases about 50%, and in winter, the resistance increases about 100% [14]. That added resistance will introduce more speed losses, and those speed losses may even accumulate and cause a severe delay in the destination. Therefore, energy management considering speed losses in uncertain wave and wind is essential for the maritime industry.

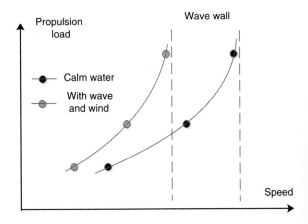

Fig. 5.3 Relationship between the propulsion load and the cruising speed

The performance of ships in wave and wind has been studied for a long time, and an empirical model is formulated in [15] by various CFD simulations, which are shown as follows.

$$R_t^T = R^C + \Delta R_t^{wave} + \Delta R_t^{wind} \tag{5.1}$$

$$\Delta R_t^{wave} = \frac{1}{\mathcal{L}} \cdot \rho_{water} \cdot g \cdot h_t^2 \cdot B_{int}^2 \cdot C^{D.wat}(\tau_t, \theta_t) \tag{5.2}$$

$$\Delta R_t^{wind} = \frac{1}{2} \cdot \rho_{air} \cdot S_{int} \cdot C^{D.air} \cdot \left[\begin{array}{c} \left(v_t^c + v_t^{wind}\cos\theta_t\right)^2 \\ -\left(v_t^c\right)^2 \end{array} \right] \tag{5.3}$$

$$v_t = \sqrt[c_2]{\frac{R^C}{R_t^T}} \cdot v_t^c \tag{5.4}$$

where R_t^T is the total resistance; R^C is the resistance of calm water; ΔR_t^{wave}, ΔR_t^{wind} are the added resistances of wave and wind; $\mathcal{L}$ is the ship length; ρ_{water} is the density of water; g is the acceleration of gravity; h_t is the wave height; B_{int} is the breath of ship intersection; $C^{D.wat}(\tau_t, \theta_t)$ is the added resistance coefficient, which is determined by wave-length τ_t and weather direction θ_t; ρ_{air} is the density of air; S_{int} is the area of ship intersection; $C^{D.air}$ is the air drag coefficient; v_t^{wind} is the wind speed; v_t^c and v_t are the cruising speed in calm water and wave/wind, respectively.

From the above Eqs. (5.1)–(5.4), there are four main decision variables to calculate the speed loss, i.e., wave height denoted as h_t, wavelength denoted as τ_t, wind speed v_t^{wind} and the weather direction θ_t. It should be noted that the weather direction is defined as the angle between the wind and the ship sailing direction. Since the wave has a similar direction with the wind, weather direction is used to indicate the influence of wave and wind.

Reference [15] has comprehensively studied the speed performance in wave and wind, and gives some fitting curves to calculate $C^{D.wat}$ under different weather direction (under B.N. 6), shown as Fig. 5.4. We can see this coefficient differs from each other when the weather direction changes.

With the above model, the speed loss under uncertain waves and wind can be predicted. Then in the energy management model, the speed loss can be considered in the voyage scheduling, and the propulsion system can response to the speed loss and ensures the punctuality of the ship's navigation.

5.2.2 Adverse Weather Conditions

Adverse weather conditions are those scenarios or areas which are not suitable for navigation [16, 17], and the ships should avoid this type of area for safety. Adverse

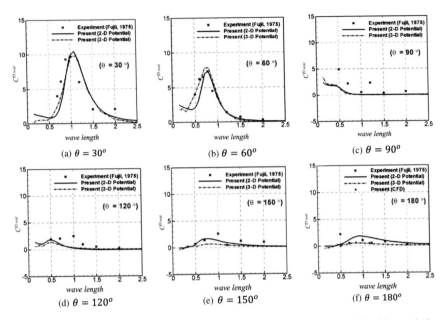

Fig. 5.4 Fitting curves to calculate added resistance of wave. Reprinted from [15], with permission from Elsevier

weather conditions generally include the typhoon or strong ocean current and the following Fig. 5.5 shows the influence of adverse weather conditions on the ship's navigation.

In Fig. 5.5, the primary navigation route is from Singapore to Inchon. The red dash line is the conventional navigation route from Singapore to Inchon due to the shortest distance. However, under pre-voyage weather forecasting, this navigation route is under the influence of a typhoon. Based on this information, the first stage chooses another navigation route (blue dash line) to keep away from the typhoon. In real-time navigation (second stage), the forecasting trajectory of typhoons may change to the black line, and the navigation route obtained in the first stage may still under the influences of typhoons. In this case, the second stage will modify the navigation route and the corresponding cruising speeds as the purple dash line for safe sailing.

From above, the uncertainties of adverse weather conditions come from the weather forecasting error, and the navigation route changes led by the adverse weather conditions will have different energy requirements on the ship power system.

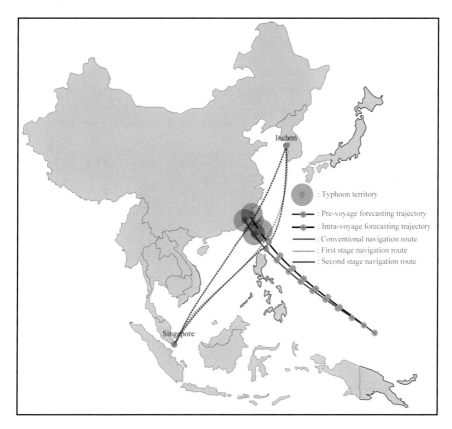

: Typhoon territory

: Pre-voyage forecasting trajectory

: Intra-voyage forecasting trajectory

: Conventional navigation route

: First stage navigation route

: Second stage navigation route

Fig. 5.5 Adverse weather conditions and the two-stage adjustment

5.2.3 Calls-for-Service Uncertainties

The former two types of uncertainties mainly influence the operation of ships and will bring delays to the destination, which brings calls-for service uncertainties to the seaport or other service facilities, such as islands or ocean platforms.

Generally, the services provided to the ships are classified as (1) the logistic services, i.e., cargo handling, and (2) the electric service, i.e., cold-ironing. Since the ships may not arrive on time for different reasons, as stated above, all the services may be delayed. Figure 5.6 shows the influences of calls-for-service uncertainties.

From above, the calls-for-service delays led to different power demand curves, which require different energy schemes. There are two main types of power demand changes, i.e., service delay and service accumulation. The service delay will not change the shape of power demand but only delays them, like the cold ironing power. The other type is the service accumulation, like cargo handling. This type of service has a constant total service workload, and if the service is delayed and the service will accumulate to increase the maximal power demand.

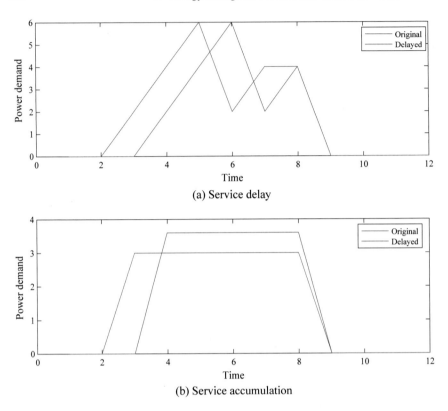

Fig. 5.6 Influences of calls-for-service

5.3 Energy Source Uncertainties

5.3.1 Renewable Energy Uncertainties

Nowadays, environmental issues have been the major concern from the globe, and renewable energy is gradually widely spread in the maritime grids, as we have stated in Chap. 1. However, renewable energy generally has high intermittency and a specified energy management method should accommodate this uncertainty. The following Fig. 5.7 gives a typical wind speed pattern.

The wind speed pattern can be depicted as a spectrum, and a high value indicates a high variation in that timescale [18]. In Fig. 5.7, the first peak is in the timescale of minutes, and the sites with high average wind speed tend to have a lower peak. This variation, referred to as the short-term variation, has been mitigated by many control strategies [19–21]. In the timescale of more than one day (Macro-meteorological range), there are three peaks, (1) Diurnal pattern, or named as the day-night pattern, which is led by the temperature difference between day and night; (2) depressions

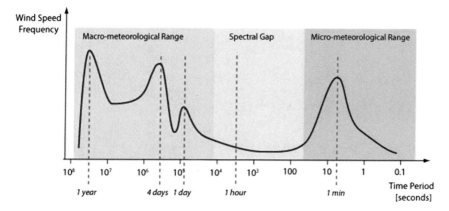

Fig. 5.7 Wind speed patterns. Reprinted from [18], open access

and anti-cyclones, and this phenomenon is more distinct in oceanic than continental regions.; (3) annual pattern, varies with the degree of latitude and vanishes close to the equator. In the following Fig. 5.8a, b, the power outputs of different wind turbines in a day and different seasons are shown.

From the above figure, we can see significant variations by different wind turbines and different seasons. As for the photovoltaic energy, the variations by different modules and different seasons are shown in the following Fig. 5.9a, b.

As above, the power outputs of the wind farm and photovoltaic farm are highly fluctuating, and even after deliberate forecasting, the error is still inevitable. Table 5.2 gives the forecasting error of renewable energy through various methods. The root-mean-square error (RMSE) are around 1–5%, which should be considered in the energy management of maritime grids.

5.3.2   Main Grid Uncertainties

The maritime grids can be mainly operated in (1) grid-connected mode; and (2) isolated mode. Two modes are shiftable for most of the maritime grids. For example, the ships are in isolated mode when sailing, and are in grid-connected mode when receiving the cold-ironing power from the seaport. For a seaport, it can also operate in isolated mode when having enough generators or renewable energy integration.

When in grid-connected mode, the main grid is generally the main energy source of maritime grids. However, there will be many uncertain failures that happened in the main grid and even cause a loss of power. The maritime grids generally don't have a strong network structure, and therefore an energy management method with considering the main grid failure is essential for the safety of maritime grids [26].

Besides, the main grid and the maritime grid maybe not under the same administrator, and the maritime grid should purchase electricity from the main grid, and

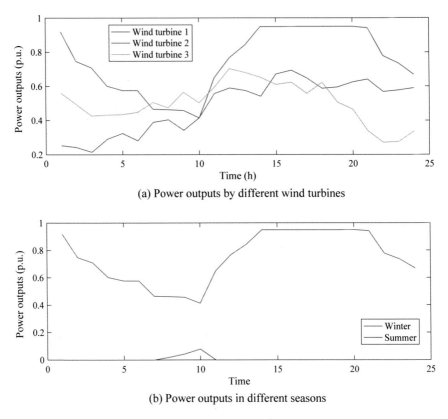

(a) Power outputs by different wind turbines

(b) Power outputs in different seasons

Fig. 5.8 Power outputs of different wind turbines and in different seasons

the electricity prices also have uncertainties. The maritime grid should aggregate the total power demand and negotiate the price with the main grid. The price may change in every round of negotiation [27], which also brings the main grid uncertainties.

5.3.3   Equipment Uncertainties

The equipment uncertainties in maritime grids come from two aspects: (1) the equipment failure; and (2) the scheduled maintenance. Their difference is the equipment failure may happen unattended but the latter one is planned.

For the equipment failure, the energy management system of maritime grids has to make enough power reserve for each severe scenario [8]. In [28, 29], to avoid the influence of the onboard generator's failure, the generation system have reserved a certain part of capacity, which are the same in ships and seaports. For a seaport,

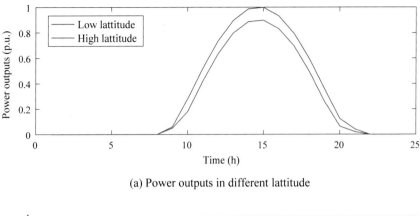

(a) Power outputs in different lattitude

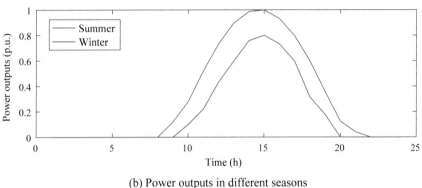

(b) Power outputs in different seasons

Fig. 5.9 Power outputs of different photovoltaic modules and in different seasons

Table 5.2 Forecasting error by different methods

Methods	Renewable energy	Timescale	Error (%)	References
f-ARIMA	Wind	Day-ahead	5.35	[22]
ANN	Wind	Day-ahead	1.32–1.56	[23]
SVM	PV	120 h	1.21	[24]
ARIMA	PV	1~39 h	21	[25]

the power reserve ratio can be lower since the main grid can provide enough power with high reliability, but the within generators still need to be standby for uncertain failure.

For the scheduled maintenance, the equipment out of service is known in advance, and the energy management system can make necessary adjustments. For example, when a generator in a seaport is planned to be in maintenance, the administrator of the seaport will give a new energy plan to the upper main grid to purchase more electricity [27].

5.4 Data-Driven Optimization with Uncertainties

5.4.1 General Model

The main types of uncertainties in the operation of maritime grids are illustrated as above. To ensure the safety and reliability of maritime grids, considering the above uncertainties in energy management is necessary. Nowadays, stochastic optimization [30–32] and robust optimization [32–34] are two main types to address the uncertainties, which are shown as following Eqs. (5.5) and (5.6), respectively.

$$\min_{x \in X} g(x) + E\left(\min_{y \in Y(x,\xi)} f(y)\right) \tag{5.5}$$

$$\min_{x \in X} g(x) + \max_{\xi \in U}\left(\min_{y \in Y(x,\xi)} f(y)\right) \tag{5.6}$$

In stochastic optimization (Eq. (5.5)), x is the first stage decision variables which are not determined by uncertainties; X is the feasible region of x; $g(x)$ is the objective function of the first stage; ξ is the uncertain variables, and $Y(x, \xi)$ is the feasible region of y determined by x and ξ; $f(y)$ is the objective function of the second stage; $E(\cdot)$ is the expectation. In this model, the uncertain variable ξ is depicted by the probability distribution, such as the probability distribution of equipment failure, or the probability distribution of renewable energy output, and so on. Then stochastic optimization seeks the optimal solution within the feasible region defined by the probability distributions.

In robust optimization (Eq. (5.6)), the main difference is the uncertain variable ξ is described by the uncertainty set U, including the upper/lower limits and the uncertainty budget, which mainly has polyhedral models [35] and ellipsoid models [36]. Then robust optimization seeks the optimal solution in the worst case in the defined uncertainty set and brings conservatism. With above, the primary problem of the uncertainty modeling is how to determine the feasible regions, such as the probability distributions in stochastic optimization and the uncertainty set in robust optimization.

As above, how to get the range of uncertain variables, i.e., the probability distribution function or the uncertainty set of ξ, is the basic problem of the optimization model. Nowadays, with the development of measurement and communication technology, more operating data can be transmitted and stored in the control center in real-time. How to use this type of massive data to model the feasible region of uncertainty has become a hot topic, and various methods have been proposed.

5.4.2 Data-Driven Stochastic Modeling

Stochastic modeling is to get the probability distribution functions of uncertain variables, and there are three types in general, (1) the non-parametric probability modeling; and (2) stochastic process modeling and (3) artificial intelligence methods.

The non-parametric probability modeling method directly extracts features from the original dataset and doesn't limit the probabilistic distribution prototype [37], thus may have higher accuracy when having limited knowledge on the dataset characteristics. Based on the diffusion-based density method, [38] proposes a non-parametric probabilistic model for wind speed. Later on, [39] proposes a model for wind speed combined the non-parametric probability modeling and auto-regression modeling. Then based on the non-parametric probability modeling, [40] formulates a probabilistic optimal economy dispatch model for a renewable integrated microgrid, and the case study proves the proposed method can improve the economic behaviors during uncertainties.

The basic idea of stochastic process modeling is to use a series of simple kernel functions to fit the complex function [41]. Based on different basis functions, stochastic process modeling has many representatives. The autoregression and moving average (ARMA) method is one of them and has been utilized in renewable power prediction, and power demand prediction [22, 25]. To reduce the dimension of the dataset, many reduction algorithms are implemented. Based on Karhunen-Loeve expansion, a time-space modeling method for renewable energy is proposed in [42, 43]. Then [44] proposes a solution method for this uncertainty modeling, and shows a lower computational burden with acceptable accuracy.

Compared with the above two types, the methods based on artificial intelligence has stronger data mining ability. The uncertain set can be directly modeled and no necessary to follow the conventional process of "probability distribution formulation-sampling-scenario reduction". Until now, various methods, such as the generative adversarial network (GAN) [45], recurrent neural network (RNN) [46], extreme learning machine (ELM) [47], are implemented to provide uncertain set by massive original dataset.

5.4.3 Data-Driven Robust Modeling

Robust modeling is to get the set of uncertain variables, and there are also three types in general, (1) the polyhedral set; and (2) the ellipsoid set and (3) the uncertain set based on scenarios.

The polyhedral set is the most commonly used uncertainty set for robust modeling, which is based on a series of upper and lower limits, shown as Eq. (5.7).

$$U = \left\{ \xi_t | \underline{\xi} \le \xi_t \le \bar{\xi}, \forall t \in T \right\} \qquad (5.7)$$

where $\underline{\xi}$ and $\bar{\xi}$ are the lower and upper limits of ξ_t. If the uncertainty series follows the Markov law, the lower and upper limits may become $\underline{\xi} = \underline{\xi}_t(\xi_{t-1})$ and $\bar{\xi} = \bar{\xi}_t(\xi_{t-1})$. To limit the range of uncertain variables, uncertainty budget constraints may be added, shown as Eq. (5.8).

$$\underline{\eta} \leq \sum_t \xi_t \Big/ \mu \cdot |T| \leq \bar{\eta} \tag{5.8}$$

where μ is the expectation of ξ; and $\underline{\eta}$, $\bar{\eta}$ are the lower and upper budgets of uncertain variable ξ. The uncertainty budget is used to limit the dramatic changes and reduce the conservatism of the robust model.

The second type is the ellipsoid set, which aims to solve the inconsistent characteristic at the boundary of the uncertain set. A general form is shown in Eqn.

$$U = \left\{ \xi_t \Big| (\xi - \mu)^T \cdot \sum(\xi - \mu)^{-1} \leq \Gamma \right\} \tag{5.9}$$

where μ is the expectation of ξ; and $\sum$ is the correlation matrix of ξ. Li et al. [48] use the ellipsoid set to model the uncertainties, and find the ellipsoid set can better represent the uncertainty when approaching the boundary. Kumar and Yildirim [49] proposes the minimum volume enclosing ellipsoid (MVEE) method to limit the uncertainty in the smallest ellipsoid and reduce the conservatism. Based on MVEE, [50] studies the robust optimization based on the ellipsoid set, and proposes an invalid constraint reduction method to simplify the solution method.

Besides the above two modeling methods, there is a modeling method based on extreme conditions. In [51], an ellipsoid set of uncertainty is first formulated and then several extreme points are selected to form a convex set. The formulated robust model is shown as follows.

$$\begin{cases} \max\limits_{\xi_n \in U_n} \left(\min\limits_{x, y_n} f(x, y_n, \xi_n) \right) \\ s.t. A(x, y_n, \xi_n) = 0 n = 1, 2, \ldots, N \\ B(x, y_n, \xi_n) \leq 0 n = 1, 2, \ldots, N \end{cases} \tag{5.10}$$

where ξ_n is the uncertain variable in the n-th extreme scenarios, and y_n is the corresponding second stage decision variables; A and B are the equality and inequality constraints, respectively.

Another robust modeling formulates the uncertain set as a convex envelope to contain all the pre-given extreme points and can be shown as Eq. (5.11). α_n is the ratio for the n-th extreme scenario.

$$U = \left\{ \xi | \xi = \sum_n \alpha_n \cdot \xi_n, \sum_n \alpha_n = 1, \alpha_n \geq 0 \right\} \tag{5.11}$$

5.5 Typical Problems

5.5.1 Energy Management for Photovoltaic (PV) Uncertainties in AES

As the main representative of maritime grids, AESs face many uncertainties during navigation. This Chapter focuses on the uncertainties of onboard photovoltaic (PV) integration. This research is illustrated in detail in [52].

(1) Onboard PV power forecasting

In land-based PV power forecasting, the PV power is determined by three factors, i.e., the irradiation density, denoted as I^{Gh}, and the angle between solar rays and the PV modules, denoted as θ, and the generation efficiency, usually determined by the ambient temperature [53], denoted as η^{PV}. However, some differences compared with the load-based applications should be incorporated into the onboard PV forecasting.

The first difference is that the ship will constantly move along the navigation route. As shown in Fig. 5.10, the ship has different locations when t_1 and t_2, meanwhile the direction of solar rays, as well as the ambient temperature along the navigation route, are also changed. Therefore, it is sensible to utilize the measured data along the route, rather than the data in a stationary place to predict the PV generation.

The second difference is that the shipboard deck will constantly swing when cruising and change the angle between solar rays and the PV modules [53], shown in Fig. 5.11. The angles between solar rays and ship decks become $(\theta \pm \phi)$, which further affects the PV generation outputs. In general, the swinging direction of ships

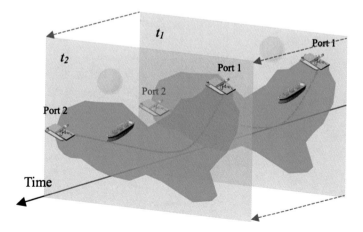

Fig. 5.10 An illustration on the moving of ships

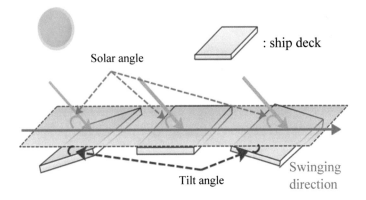

Solar angle

: ship deck

Tilt angle

Swinging direction

Fig. 5.11 Definition of the angle of solar ray and the tilt angle. Reprinted from [52], with permission from IEEE

is the same with the wind direction and the tilt angle is determined by the wind speed. So, it is necessary to incorporate wind speed along the navigation route to forecast the tilt angle range of ships.

(2)   Two-stage robust modeling framework

The above two characteristics are both considered, and this Chapter proposes a data-driven PV generation uncertainty characterization method, shown as the below Fig. 5.12a. The general framework of the two-stage robust modeling is shown as Fig. 5.12b.

In Fig. 5.12a, owing to the high scalability and fast computational speed, the Extreme Learning Machine (ELM) is regarded as a useful learning technique for training a single hidden-stage feed-forward neural network [54]. In Fig. 5.12b, the forecasting values and error of irradiation density, wind speed, and temperature

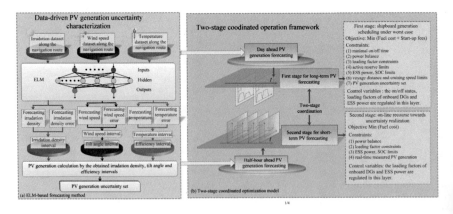

Fig. 5.12 Overall framework of proposed model. Reprinted from [52], with permission from IEEE

are obtained by ELM. Then three intervals, i.e., the irradiation density intervals $\left[I_{min,t}^{Gh}, I_{max,t}^{Gh}\right]$, the tilt angle intervals $\left[\phi_t^{min}, \phi_t^{max}\right]$ and the PV generation efficiency intervals $\left[\eta_{min,t}^{PV}, \eta_{max,t}^{PV}\right]$ are obtained by two different ways, i.e., $\left[I_{min,t}^{Gh}, I_{max,t}^{Gh}\right]$ is calculated by the forecasting values and error, and $\left[\phi_t^{min}, \phi_t^{max}\right]$, $\left[\eta_{min,t}^{PV}, \eta_{max,t}^{PV}\right]$ are calculated by the forecasting wind speed intervals and temperature intervals, since higher wind speed and temperature will lead to larger rolling effect and generation efficiency, respectively.

$$P_t^{PV} = \eta_t^{PV} \cdot A_{PV} \cdot I_t^{Gh} \cdot \left[\begin{array}{c} \cos\theta_t + C_{\phi_1}\left(\cos\frac{\phi_t}{2}\right)^2 \\ +C_{\phi_2}\left(\sin\frac{\phi_t}{2}\right)^2 \end{array} \right] \qquad (5.12)$$

Based on the obtained uncertain PV generation as (5.12), the proposed two-stage multi-timescale coordinated operation framework aims to coordinate different controllable resources in different timescales according to their different response characteristics considering the uncertain PV outputs, which is shown in Fig. 5.12b. In the day-ahead time-window, i.e., the first stage, the DGs' on/off states and the cruising speed, which cannot instantly respond to the uncertainties, are optimized based on day-ahead interval predictions of the PV generation. This stage aims to dispatch the DGs and ESS on a large time horizon to fulfill propulsion and service loads in the worst case of PV generation.

During the half-hour-ahead online operation time-window, i.e., the second stage, the loading factor of DGs and ESS are re-dispatched based on half-hour-ahead predictions of the PV generation. The half-hour-ahead predictions tend to be more accurate and they can be regarded as the uncertainty realization. Thus, the second-stage operation aims to compensate for the first-stage operation when the uncertainties realize in practice.

(3) Case description

In this study, a typical medium voltage direct current (MVDC) 4-DGs AES case is used to verify the proposed method. The topology and navigation data of this 4-DG AES are shown in Figs. 5.13 and 5.14, respectively. The topology is from [55], which follows the ABS-R2 standard [56]. In Fig. 5.13, 4 DGs are connected in two buses via AC/DC converters, and the circuit breaker is normally open. In general cases, two buses are located in different watertight compartments for avoiding operating risk. As for the PV generation uncertainty set characterization, the training datasets are also applied to [53], which are deduced from real-world navigation from Dalian, China to Aden, Yemen, and 2 MW PV modules are integrated into the AES for future applications. Other detailed parameters can be found in [52].

(4) Case study

To test the validity of the proposed forecasting process, three forecasting methods are compared. The results are shown in Fig. 5.15.

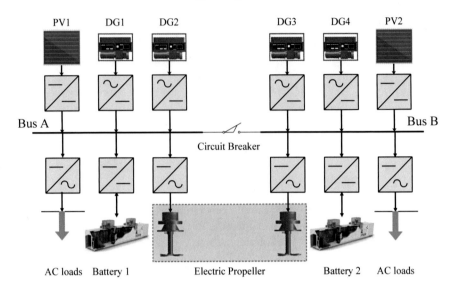

Fig. 5.13   Topology of 4-DG AES. Reprinted from [52], with permission from IEEE

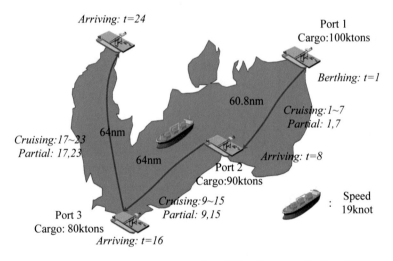

Fig. 5.14   Navigation scheme of AES. Reprinted from [53], with permission from IEEE

Forecasting method A: the proposed method considering both the movement and tilt angle (wind speed);

Forecasting method B: the proposed method without considering the tilt angle (wind speed);

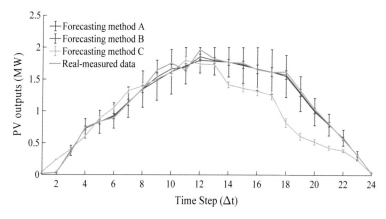

Fig. 5.15 Forecasting results under different methods. Reprinted from [53], with permission from IEEE

Forecasting method C: forecasting method only using the dataset in a stationary place (irradiation density, temperature) without considering the tilt angle.

From Fig. 5.15, the following conclusions can be found, (1) from the comparison between methods A and B, the forecasting intervals become much wider when considering tilt angle. This phenomenon suggests the rolling of the shipboard deck will bring more uncertainties to the PV generation, and if it is ignored, an optimistic scheme may be obtained; (2) from the comparison between method B and C, the forecast error of method C becomes rather large when the ship is away from the initial port ($t = 14 \sim 24$), which suggests the necessity to use the dataset along the navigation route to predict the PV generation.

The energy scheduling schemes in two stages are shown in Figs. 5.16 and 5.17, respectively. From Fig. 5.16, since the PV generations in the second stage are all larger than the worst case, the DGs' outputs are further replaced by the PV integration, which introduces further FC reductions. From Fig. 5.17, the ESS power in most of the partial intervals is increased, which means the PV generation increments are directly charged to the ESS in the partial intervals, therefore in the cruising intervals the ESS has more energy to shed the power demands of DGs than the first stage.

The above results manifest that, since the worst case of PV generation only happens in a small probability, therefore the single-stage robust method will introduce plenty of conservatism to the operating scheme, which leads to wastes on the PV generation. In this section, the online half-hour-ahead operation can effectively go against the uncertainty realization to improve the overall energy utilization efficiency while satisfying the constraints.

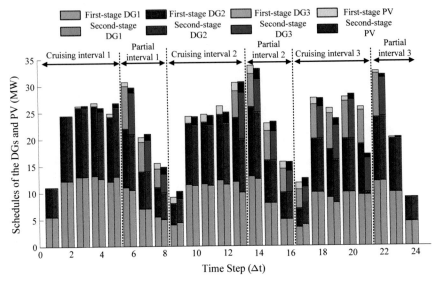

Fig. 5.16 Scheduling schemes of DGs in the first and second stages. Reprinted from [53], with permission from IEEE

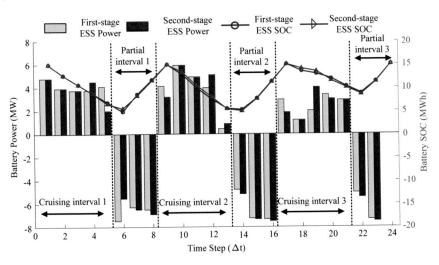

Fig. 5.17 ESS outputs and SOC in the first and second stages. Reprinted from [52], with permission from IEEE

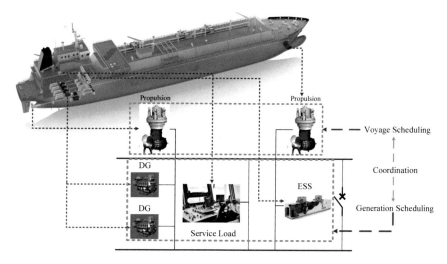

Fig. 5.18 Coordinated generation-voyage scheduling. Reprinted from [57], with permission from Elsevier

5.5.2 Energy Management for Navigation Uncertainties in AES

(1) Problem formulation

Besides the above PV power uncertainties, the navigation uncertainties are also commonly faced during the operation of AES. Fang and Xu [57] has studied this problem in detail, which is illustrated below. As shown in Fig. 5.18, the shipboard microgrid of an AES consists of DGs and ESS to meet the propulsion and service loads.

Compared with the conventional land-based microgrids, the AES (mobile microgrid) has the total voyage distance to the ports as a mandatory requirement, and therefore put extra constraints on the cruising speed of AES, as well as on the propulsion load. The generation scheduling aims to an economic energy scheme and the voyage scheduling aims to a punctual energy scheme. Both of them consist of a coordinated generation-voyage scheduling problem.

The speed loss when considering navigation uncertainties can be calculated by (5.1)–(5.4). The uncertainty set of the proposed model is formulated as following (4.13).

In this section, the wave height h, wavelength τ, weather direction angle θ and wind speed v^{wind} are four uncertain variables. In (4.13), $\overline{h_t}$, $\overline{\tau_t}$ and $\overline{v_t^{wind}}$ are the expectations of corresponding uncertain variables; $\underline{\mu}$ and $\bar{\mu}$ are the lower and upper budgets of the uncertainty set, when the lower budget falls and the upper budget rises, it means that the uncertainty set can cover higher uncertainty, leading to a higher robustness degree. Then the robust model shown in (5.6) is utilized to consider the worst influence by the navigation uncertainties.

$$
\mathcal{U} = \left\{
\begin{array}{l}
h_t \in \mathbb{R}^{|T|} : h^{min} \leq h_t \leq h^{max}, \forall t \in T \\[6pt]
\mu^h \leq \sum_{t \in T} h_t \Big/ \sum_{t \in T} \overline{h_t} \leq \overline{\mu^h} \\[6pt]
\tau_t \in \mathbb{R}^{|T|} : \tau^{min} \leq \tau_t \leq \tau^{max}, \forall t \in T \\[6pt]
\mu^\tau \leq \sum_{t \in T} \tau_t \Big/ \sum_{t \in T} \overline{\tau_t} \leq \overline{\mu^\tau} \\[6pt]
v_t^{wind} \in \mathbb{R}^{|T|} : v_{min}^{wind} \leq v_t^{wind} \leq v_{max}^{wind} \\[6pt]
\mu^v \leq \sum_{t \in T} v_t^{wind} \Big/ \sum_{t \in T} \overline{v_t^{wind}} \leq \overline{\mu^v} \\[6pt]
\vartheta \in [0, 180^o] : \theta_t = \varsigma \cdot \vartheta, \forall \varsigma \leq 180/\vartheta \in \mathbb{N}
\end{array}
\right\}
\tag{5.13}
$$

(2) Case study

To test the effects of proposed robust model on the on-time rates, 500 water current scenarios are randomly sampled according to uniform distributions in each time-interval, denoted as $(h_{l,t}, \tau_{l,t}, v_{l,t}^{wind})$, $t = 1 \sim 24$, $i = 1 \sim 500$. Robust 1 (The formulated robust model considering navigation uncertainties, abbreviated as R1) and Non-robust (conventional coordinated generation-voyage scheduling without navigation uncertainties, abbreviated as NR1) are set as operating strategies, respectively. The corresponding voyage distances of each sample at the scheduled time under $\theta = 30^o$ are shown in Fig. 5.19. The cruising speed and EEOI are shown in Fig. 5.20. The generation scheduling schemes are shown in Fig. 5.21. The worst speed loss and the corresponding on-time rates under different θ with or without wind are shown in Fig. 5.22.

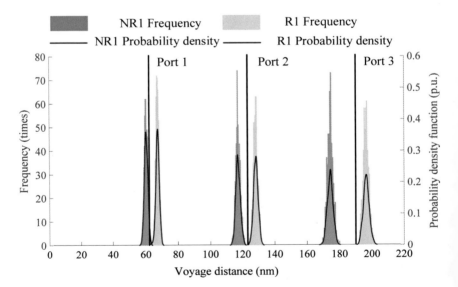

Fig. 5.19 On-time rates of different voyage schedules. Reprinted from [57], with permission from Elsevier

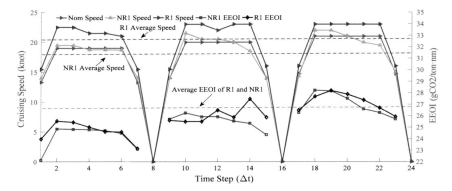

Fig. 5.20 Comparisons between NR1 and R1. Reprinted from [57], with permission from Elsevier

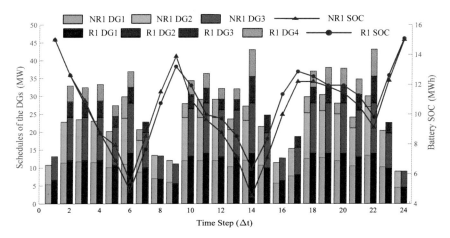

Fig. 5.21 Scheduling schemes of DGs and ESS of NR1 and R1. Reprinted from [57], with permission from Elsevier

Figure 5.19 clearly shows that the influences of uncertain water and wind will constantly accumulate during the voyage, which leads to an average 13 nm delay, leading to a 0% on-time rate of NR1 at the terminal port. However, the proposed robust model can accommodate these uncertainties by adjusting the outputs of the DGs and ESS. Accordingly, the corresponding on-time rates of R1 to each port are all 100%.

The reason to ensure the on-time rates of proposed method can be inferred from Figs. 5.20 and 5.21. The first berthing time-interval, $t = 0$, is not included in the analysis since the cruising speed and corresponding propulsion load are both zeros.

In Fig. 5.20, the cruising speed of robust model is higher than non-robust model, so able to cover the speed loss led by the wave and wind. Higher cruising speed suggests heavier propulsion load, so the corresponding outputs of DGs are all increased to meet the power demand increments, which leads to a higher FC. Specifically, in

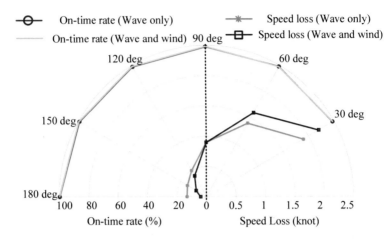

Fig. 5.22 Worst speed loss and corresponding on-time rates. Reprinted from [57], with permission from Elsevier

Fig. 5.21, NR1 uses no more than 3 DGs all the time, even 2 DGs in $t = 1 \sim 5, 8, 9, 15 \sim 17, 23, 24$. Correspondingly, R1 uses 4 DGs in most time during the voyage, only except the partial speed time-intervals, $t = 1, 7 \sim 9, 15 \sim 17, 23, 24$.

Figure 5.22 shows the worst speed loss and corresponding on-time rates. The yellow and red curves show that the proposed method can ensure a 100% on-time rate for all uncertain scenarios. The green curves show that the water wave always has negative impacts on the cruising speed, but the effect will gradually fade with the increment of the weather direction angle, which leads to the speed loss reductions.

Besides, it can be observed from Fig. 5.22 that, unlike water wave, the wind has quite different impacts on the speed loss in different scenarios, e.g. when $\theta \in [30°, 90°]$, the wind will increases speed loss, while when $\theta \in [90°, 180°]$, the wind can reduce speed loss. Especially when $\theta = 150°$ and $180°$, the speed loss under wave and wind are less than 0.5knot, thereby its negative impacts on the voyage scheduling can be greatly reduced. This is also the key reason for the cruising ships to choose their navigation route to the leeward side of wind.

References

1. Kim, M.: Estimation of added resistance and ship speed loss in a seaway. Ocean Eng. **141**, 465–476 (2017)
2. Harvald, S.A.: Resistance and propulsion of ships. Krieger Publishing Company, P.O. Box 9542, Melbourne, FL United States (1992)
3. Shao, W., Zhou, P., Thong, S.K.: Development of a novel forward dynamic programming method for weather routing. J. Mar. Sci. Technol. **17**(2), 239–251 (2012)
4. Padhy, C.P., Sen, D., Bhaskaran, P.K.: Application of wave model for weather routing of ships in the North Indian Ocean. Nat. Hazards **44**(3), 373–385 (2008)

5. Lin, Y.H., Fang, M.C., Yeung, R.W.: The optimization of ship weather-routing algorithm based on the composite influence of multi-dynamic elements. Appl. Ocean Res. **43**, 184–194 (2013)
6. Krata, P., Szlapczynska, J.: Ship weather routing optimization with dynamic constraints based on reliable synchronous roll prediction. Ocean Eng. **150**, 124–137 (2018)
7. Lai, K., Illindala, M.: Graph theory based shipboard power system expansion strategy for enhanced resilience. IEEE Trans. Ind. Appl. **54**(6), 5691–5699 (2018)
8. Xu, Q., Yang, B., Han, Q., et al.: Optimal power management for failure mode of MVDC microgrids in all-electric ships. IEEE Trans. Power Syst. **34**(2), 1054–1067 (2018)
9. Louit, D., Pascual, R., Banjevic, D., et al.: Optimization models for critical spare parts inventories—a reliability approach. J. Oper. Res. Soc. **62**(6), 992–1004 (2011)
10. Evans, P.M., Schultz, R.D.: Rotating electrical machine with electromagnetic and permanent magnet excitation. U.S. Patent 5,663,605. 1997-9-2
11. Sebille, D., Westerholt, E.: Control device for a reversible rotating electrical machine. U.S. Patent 7,102,304. 2006-9-5
12. Iris, Ç., Lam, J.S.: A review of energy efficiency in ports: Operational strategies, technologies and energy management systems. Renew. Sustain. Energy Rev. **112**, 170–182 (2019)
13. Jurong Port starts world's largest port-based solar facility. https://www.businesstimes.com.sg/energy-commodities/jurong-port-starts-worlds-largest-port-based-solar-facility
14. Man Diesel & Turbo, Basic principal of ship propulsion, 2012, Technical Report
15. Kim, M., Hizir, O., Turan, O., et al.: Estimation of added resistance and ship speed loss in a seaway. Ocean Eng. **141**, 465–476 (2017)
16. Sumalee, A., Uchida, K., Lam, W.: Stochastic multi-modal transport network under demand uncertainties and adverse weather condition. Transp. Res. Part C: Emerg. Technol. **19**(2), 338–350 (2011)
17. Sprenger, F., Maron, A., Delefortrie, G., et al.: Experimental studies on seakeeping and maneuverability of ships in adverse weather conditions. J. Ship Res. **61**(3), 131–152 (2017)
18. Hoven, V.: Power spectrum of horizontal wind speed in the frequency range from 0.0007 to 900 cycles per hour. J. Meteorol. **14**(2), 160–164 (1957)
19. Yao, J., Li, H., Liao, Y., et al.: An improved control strategy of limiting the DC-link voltage fluctuation for a doubly fed induction wind generator. IEEE Trans. Power Electron. **23**(3), 1205–1213 (2008)
20. Jiang, Q., Gong, Y., Wang, H.: A battery energy storage system dual-layer control strategy for mitigating wind farm fluctuations. IEEE Trans. Power Syst. **28**(3), 3263–3273 (2013)
21. Li, X., Hui, D., Lai, X.: Battery energy storage station (BESS)-based smoothing control of photovoltaic (PV) and wind power generation fluctuations. IEEE Trans. Sustain. Energy **4**(2), 464–473 (2013)
22. Kavasseri, R., Seetharaman, K.: Day-ahead wind speed forecasting using f-ARIMA models. Renew. Energy **34**(5), 1388–1393 (2009)
23. Li, G., Shi, J.: On comparing three artificial neural networks for wind speed forecasting. Appl. Energy **87**(7), 2313–2320 (2010)
24. Shi, J., Lee, W., Liu, Y., et al.: Forecasting power output of photovoltaic systems based on weather classification and support vector machines. IEEE Trans. Ind. Appl. **48**(3), 1064–1069 (2012)
25. Fernandez-Jimenez, L., Muñoz-Jimenez, A., Falces, A., et al.: Short-term power forecasting system for photovoltaic plants. Renew. Energy **44**, 311–317 (2012)
26. Li, J., Liu, Y., Wu, L.: Optimal operation for community-based multi-party microgrid in grid-connected and islanded modes. IEEE Trans. Smart Grid **9**(2), 756–765 (2016)
27. Kanellos, F.D., Volanis, E., Hatziargyriou, N.D.: Power management method for large ports with multi-agent systems. IEEE Trans. Smart Grid **10**(2), 1259–1268 (2017)
28. Moya, O.: A spinning reserve, load shedding, and economic dispatch solution by Bender's decomposition. IEEE Trans. Power Syst. **20**(1), 384–388 (2005)
29. Jaefari-Nokandi, M., Monsef, H.: Scheduling of spinning reserve considering customer choice on reliability. IEEE Trans. Power Syst. **24**(4), 1780–1789 (2009)

30. Banzo, M., Ramos, A.: Stochastic optimization model for electric power system planning of offshore wind farms. IEEE Trans. Power Syst. **26**(3), 1338–1348 (2010)
31. Abbey, C., Joós, G.: A stochastic optimization approach to rating of energy storage systems in wind-diesel isolated grids. IEEE Trans. Power Syst. **24**(1), 418–426 (2008)
32. Alqurashi, A., Etemadi, A., Khodaei, A.A.: Treatment of uncertainty for next generation power systems: State-of-the-art in stochastic optimization. Electric Power Syst. Res., **141**, 233–245 (2016)
33. Peng, C., Xie, P., Pan, L., et al.: Flexible robust optimization dispatch for hybrid wind/photovoltaic/hydro/thermal power system. IEEE Trans. Smart Grid **7**(2), 751–762 (2015)
34. Bertsimas, D., Litvinov, E., Sun, X., et al.: Adaptive robust optimization for the security constrained unit commitment problem. IEEE Trans. Power Syst. **28**(1), 52–63 (2012)
35. Minoux, M.: On robust maximum flow with polyhedral uncertainty sets. Optim. Lett. **3**(3), 367–376 (2009)
36. Ben-Tal, A., Nemirovski, A.: Robust solutions of uncertain linear programs. Oper. Res. Lett. **25**(1), 1–13 (1999)
37. Epanechnikov, V.: Non-parametric estimation of a multivariate probability density. Theory Probab. Appl. **14**(1), 153–158 (1969)
38. Xu, X., Yan, Z., Xu, S.: Estimating wind speed probability distribution by diffusion-based kernel density method. Electr. Power Syst. Res. **121**, 28–37 (2015)
39. Yuan, Y., Wang, J., Zhou, K., et al.: Monthly unit commitment model coordinated short-term scheduling and efficient solving method for renewable energy power system. Proc. CSEE **39**(18), 5336–5345 (2019)
40. Akhavan-hejazi, H., Mohsenian-rad, H.: Energy storage planning in active distribution grids: a chanceconstrained optimization with non-parametric probability functions. IEEE Trans. Smart Grid **9**(3), 1972–1985 (2016)
41. Kulkarni, V.: Modeling and Analysis of Stochastic Systems. CRC Press (2016)
42. Safta, C., Chen, R., Najm, H., et al.: Efficient uncertainty quantification in stochastic economic dispatch. IEEE Trans. Power Syst. **32**(4), 2535–2546 (2016)
43. Li, J., Ou, N., Lin, G., et al.: Compressive sensing based stochastic economic dispatch with high penetration renewables. IEEE Trans. Power Syst. **34**(2), 1438–1449 (2018)
44. Hu, Z., Xu, Y., Korkali, M., et al.: Uncertainty quantification in stochastic economic dispatch using Gaussian process emulation. 2020 IEEE Power & Energy Society Innovative Smart Grid Technologies Conference (ISGT), pp. 1–5. IEEE (2020)
45. Chen, Y., Wang, Y., Kirschen, D., et al.: Model-free renewable scenario generation using generative adversarial networks. IEEE Trans. Power Syst. **33**(3), 3265–3275 (2018)
46. Kong, W., Dong, Z., Jia, Y., et al.: Short-term residential load forecasting based on LSTM recurrent neural network. IEEE Trans. Smart Grid **10**(1), 841–851 (2017)
47. Hossain, M., Mekhilef, S., Danesh, M., et al.: Application of extreme learning machine for short term output power forecasting of three grid-connected PV systems. J. Clean. Prod. **167**, 395–405 (2017)
48. Li, P., Guan, X., Wu, J., et al.: Modeling dynamic spatial correlations of geographically distributed wind farms and constructing ellipsoidal uncertainty sets for optimization-based generation scheduling. IEEE Trans. Sustain. Energy **6**(4), 1594–1605 (2015)
49. Kumar, P., Yildirim, E.: Minimum-volume enclosing ellipsoids and core sets. J. Optim. Theory Appl. **126**(1), 1–21 (2005)
50. Ding, T., Lv, J., Bo, R., et al.: Lift-and-project MVEE based convex hull for robust SCED with wind power integration using historical data-driven modeling approach. Renew. Energy **92**, 415–427 (2016)
51. Velloso, A., Street, A., Pozo, D., et al.: Two-stage robust unit commitment for co-optimized electricity markets: An adaptive data-driven approach for scenario-based uncertainty sets. IEEE Trans. Sustain. Energy **11**(2), 958–969 (2019)
52. Fang, S., Xu, Y., Wen, S., et al.: Data-driven robust coordination of generation and demand-side in photovoltaic integrated all-electric ship microgrids. IEEE Trans. Power Syst. **35**(3), 1783–1795 (2019)

53. Wen, S., Lan, H., Hong, Y.Y., et al.: Allocation of ESS by interval optimization method considering impact of ship swinging on hybrid PV/diesel ship power system. Appl. Energy **175**, 158–167 (2016)
54. Wan, C., Lin, J., Song, Y., et al.: Probabilistic forecasting of photovoltaic generation: An efficient statistical approach. IEEE Trans. Power Syst. **32**(3), 2471–2472 (2016)
55. Kanellos, F.D., Tsekouras, G.J., Prousalidis, J.: Onboard DC grid employing smart grid technology: challenges, state of the art and future prospects. IET Electr. Syst. Transp. **5**(1), 1–11 (2014)
56. Patel, M.R.: Shipboard Electrical POWER Systems. CRC Press (2011)
57. Fang, S., Xu, Y.: Multi-objective robust energy management for all-electric shipboard microgrid under uncertain wind and wave. Int. J. Electr. Power Energy Syst. **117**, 105600 (2020)

Chapter 6
Energy Storage Management of Maritime Grids

6.1 Introduction to Energy Storage Technologies

Energy is an essential commodity and a key element for global development, and generally comes from various sources and can be mainly classified as two types, (1) the primary forms of energy and (2) the secondary forms of energy. The primary forms of energy are those energy sources that only involve extraction or capture, and the energy directly comes from nature. Typical examples are crude oil, coal, various renewable energy, natural uranium, and falling or flowing water. On the other hand, the secondary forms of energy include all the energy forms after the transformation from the primary forms of energy. The relationship between the primary forms and the secondary forms are shown in Fig. 6.1.

Secondary energy forms are generally more convenient to use and usually viewed as "energy carriers", including various types of petroleum, diesel, and electricity. The transformation technologies include oil refinery, thermal power plants, nuclear power plants, solar power plants, and so on. Among all the secondary forms of energy, electricity is the main "energy carrier" for daily lives, and power system is the corresponding man-made network to generate, transmit, and distribute electricity. Conventionally, the generation-side and demand-side of power system should be equal all the time since the electricity cannot be stored. Nowadays, with large scale of energy storage, power system will have more flexibility since energy storage can change its roles between the generation-side and the demand-side.

As a special type of power system, maritime grids also complete similar roles of "generate-transmit-distribute" as conventional power systems. For example, a seaport microgrid purchases electricity from the upper grid, and the electricity transmits to the seaport via the main substation and then distributes to different equipment within the seaport. Similarly in ships, the main and auxiliary engines generate electricity and

S. Fang and H. Wang, *Optimization-Based Energy Management for Multi-energy Maritime Grids*, Springer Series on Naval Architecture, Marine Engineering, Shipbuilding and Shipping 11, https://doi.org/10.1007/978-981-33-6734-0_6

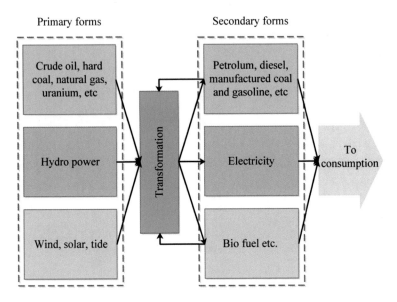

Fig. 6.1 Primary and secondary energy [1]

the electricity is transmitted and distributed by the shipboard microgrid to various load demands. In this sense, energy storage also plays an essential role to facilitate the optimal operation of maritime grids.

For ships, in [2, 3], energy storage is coordinated with the propulsion system of an AES to achieve better economic and environmental targets. Then in [4], energy storage is used to supply the energy consumption of the shipboard gas capture system. In short-term timescale, [5–7] use energy storage to mitigate propulsion fluctuations. For seaports, [8–10] classify the energy storage as an individual agent and has its energy plans to participate in the seaport operation. Molavi et al. [11] uses energy storage to facilitate renewable energy integration. Later on, [12, 13] use energy storage to recover the energy when the lifting-down of port cranes. The above literature has clearly shown that energy storage has already been an important device in maritime grids, and proper management is essential for maritime grids.

This Chapter focuses on this topic and is organized as follows. Section 6.2 gives the characteristics of different energy storage technologies, and Sect. 6.3 gives several application cases of energy storage in maritime grids. At last, Sect. 6.4 analyzes two typical problems to demonstrate the effects of energy storage management.

6.2 Characteristics of Different Energy Storage Technologies

6.2.1 Classifications of Current Energy Storage Technologies

In this section, Fig. 1.13 is re-drawn here to show the classifications of energy storage and denoted as Fig. 6.2. This Chapter focuses on conventional energy storage technologies and fuel cell will be discussed in detail in Chap. 8. The nomenclature of various energy storage technologies is shown in Table 6.1.

In the following Table 6.2, the characteristics of different energy storage are given. Since the different characteristics, we can find that different energy storage has

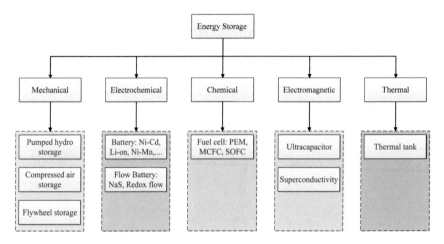

Fig. 6.2 Classification of energy storage

Table 6.1 Nomenclature of different energy storage technologies

BES: Battery Energy Storage	ZBB: Zinc-bromine flow battery
CAES: Compressed Air Energy Storage	NaS: Sodium-sulfur
FBES: Flow Battery Energy Storage	Ni-Cd: Nickel-cadmium
FESS: Flywheel Energy Storage	PSB: Polysulfide bromide battery
Li-ion: Lithium-ion	PHS: Pumped hydro storage
SMES: Superconducting magnetic energy storage	PEM: Proton exchange membrane
SCES: Supercapacitor energy storage	VRB: Vanadium redox battery

Table 6.2 Characteristics of different energy storage [14, 15]

Technologies	Investment (US$/kWh)	Energy rating (MWh)	Power rating (MW)	Specific energy (kWh/kg)	Specific power (kW/kg)
PHS	10–15	500–8000	10–1000	–	–
CAES	2–4	580, 2860	50–300	3.2–5.5	–
VRB	600	1.2–60	0.2–10	25–35	166
ZBB	500	0.1–4	0.1–1	70–90	45
PSB	450	0.005–120	0.1–15	–	–
NaS	170–200	0.4–244.8	0.05–34	100	90–230
Lead-Acid	50–100	0.001–40	0.05–10	30–50	180–200
Ni-Cd	400–2400	6.75	45	30–80	100–150
Li-ion	900–1300	0.001–50	0.01–50	80–200	200–2000
SMES	200–300	0.015	1–100	10–70	400–2000
FESS	400–800	0.025–5	0.1–20	5–100	10000+
SCES	100–300	0.01	0.05–0.2	5–15	10000+

quite different application scenarios. In the following context, some energy storage technologies which are used in maritime grids are described in detail to show their applications.

6.2.2 Battery

Among current energy storage technologies, the battery is one of the most common technologies available on the market. The battery stores energy in the electrochemical form and the battery cells are connected in series or in parallel or both to make up the desired voltage and capacity. A typical battery packs' structure is shown as Fig. 6.3, and each battery cell consists of two electrodes and an electrolyte, which are sealed in a container and then integrated into the external grid or load.

In the last decade, the technologies of battery have become much more mature, such as the lead-acid battery, nickel-cadmium battery, lithium-ion battery. Especially for lead-acid batteries, which have been researched for over 140 years and is the most mature battery technology now. Currently, tremendous efforts have been carried out to turn technologies like nickel-cadmium and lithium-ion batteries into cost-effective options for higher power applications, and their lifetimes are also important research topics.

Fig. 6.3 Illustration of battery energy storage packs

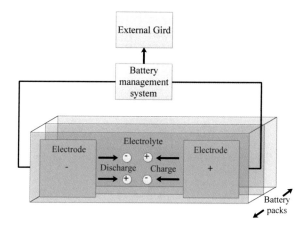

6.2.3 *Flywheel*

FES stores energy as the form of kinetic energy in a rotating mass or rotor. The stored energy is proportional to the rotor mass, location of the mass, and the rotor's rotational speed. When FES charges, it absorbs the energy from outside and accelerates the rotating speed of mass. On the other side, when the flywheel discharges, the rotating mass drives a generator to produce electrical power, and the rotating speed slows down. An illustration of flywheel energy storage is shown in Fig. 6.4.

Compared with other types of energy storage, FES can quickly respond to the power demand, and therefore be widely used in improving the power quality, load demand peak shaving, power factor correction, and load leveling. Other applications of flywheels include UPS [16], frequency response [17], smoothing wind power [18], and heavy haul locomotives [19].

The advantages of FES can be illustrated as it provides intermediate characteristics in terms of power and energy density compared with batteries and super-capacitor,

Fig. 6.4 Illustration of flywheel energy storage

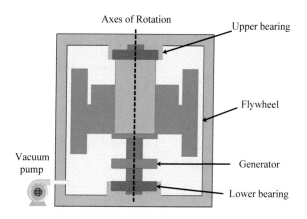

i.e., the FES has much higher power density than batteries and much higher energy density than supercapacitors. Besides, FES also caters with many shortcomings of prior energy storage technologies, i.e., less sensitivity to temperature, chemical hazardless, higher life cycle, reduced space, and weight, which is suitable for many applications. But the FES also has its shortcoming, i.e., the complex maintenance process for rotating mass.

6.2.4 Ultracapacitor

Capacitors store energy in the electric field and have a quite low equivalent series-resistance that enable them to supply the power efficiently. Generally, the capacitors are used in higher power demand scenarios, including the compensation of reactive power, mitigation of load fluctuations, and power quality issues. Capacitors usually can be classified as super-capacitors, electrolytic capacitors, and electro-static capacitors. Figure 6.5 illustrates the typical structure of a super-capacitor. The main advantages of super-capacitors are higher power density, faster charging and discharging, longer life cycles compared with other energy storage technologies. The disadvantages are the low voltage of each cell, and much higher investment cost per Watt-hour, i.e., more than 10 times compared with a lithium battery. Other drawbacks of super-capacitor include relatively low energy density, linear discharge voltage, and high self-discharge.

Fig. 6.5 Illustration of super-capacitor energy storage

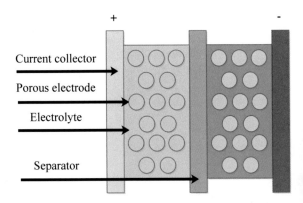

6.3 Applications of Energy Storage in Maritime Grids

6.3.1 Roles of Energy Storage in Maritime Grids

Generally, energy storage in maritime grids has three main applications, (1) as the main energy source, and (2) for long-term load leveling, shifting or shaving; and (3) for short-term power balancing.

Using energy storage as the main energy source is a recent trend for some short-trip ferries or cargo ships. Such as the first all-electric ferry "ampere" in North Europe [20], and China's first all-electric cargo ship "puffer" in 2019 [21]. Until now, there are more than 50 ships using energy storage as the main energy source in Europe. The biggest capacity is 4.16 MWh (Li-ion), the smallest capacity is 0.02 MWh (Lead-acid). The all-electric ships are about to develop in China and there will be more ships launched in the future. The advantage of using energy storage as the main energy source is nearly zero-emission, but the disadvantage is also obvious, i.e., the capacity of current energy storage technologies is limited to individually sustain a large ship for a long-distance voyage. Similar in seaports and other ocean platforms, the capacity of current energy storage is just enough to serve as auxiliary equipment. In this sense, the main application scenarios of energy storage are still in the long-term load leveling and short-term power balancing.

For the long-term load leveling, the energy storage should have enough energy density to sustain a long-time discharging. Battery is generally the main equipment to undertake this task. Nowadays, many maritime grids have installed energy storage as essential auxiliary equipment for better system characteristics. Two recent examples in China are provided as following Fig. 6.6.

The first example is the emergency supporting ship launched on April, 28th, 2020 in Shenzhen [22]. This ship has a length of 78 m and 12.8 m breadth. The deadweight is 1450 tons. The propulsion system has three diesel generators (3 × 2080 kW) and

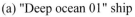

(a) "Deep ocean 01" ship   (b) Lianyungang port

Fig. 6.6 Two cases for energy storage integration in maritime grids

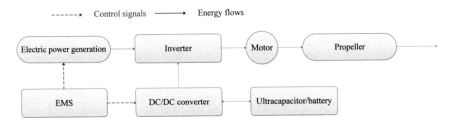

Fig. 6.7 Schematic of an electric propulsion system with ultracapacitor

two Li-ion batteries (2 × 750 kW). The second example is in Lianyungang Port which plans a battery installment (1 MW ultracapacitor + 4 MW Li-ion battery) for cold-ironing services [23]. The above two examples are both using energy storage for long-term load-leveling (hours or even longer).

For short-term power balancing, energy storage should have enough power density. This task is usually undertaken by the ultracapacitor [5] or flywheel [6], since they have enough power density and can quickly respond to the power fluctuations. Jiang et al. [24] gives a schematic of electric propulsion system with ultracapacitor, which is shown as Fig. 6.7.

In Fig. 6.7, the EMS sends control signals to the electric power generation and DC/DC converter to determine their power outputs. Then the electric power generation and ultracapacitor are both used to supply the propeller.

The applications of energy storage in maritime grids are briefly described above. To further clarify the applications, three scenarios are selected and analyzed in detail, i.e., navigation uncertainties and demand response, renewable energy integration, and energy recovery.

6.3.2 Navigation Uncertainties and Demand Response

Chapters 3 and 4 have discussed the influences of navigation uncertainties on the maritime grids. To mitigate these uncertainties, maritime grids should reserve a certain "sea margin" or "spinning margin" which can quickly respond [25]. For a maritime grid, the influences of navigation uncertainties can be described as the changes in load demands. Figure 6.8 gives an example of how energy storage mitigates the navigation uncertainties.

From the following Fig. 6.8a, the total power demand has two peaks. The main energy source need to suffer fast ramping-ups/ramping-downs, or frequent shut-downs/start-ups to follow the power demand. The influences of navigation uncertainties are similar to Fig. 6.8a, i.e., leading to many peak loads. When integrating energy storage, the main energy source and energy storage can share the total power demand, shown as Fig. 6.8b. The charging/discharging of energy storage can smooth the power demand and make the main energy source working in a steady-state, and

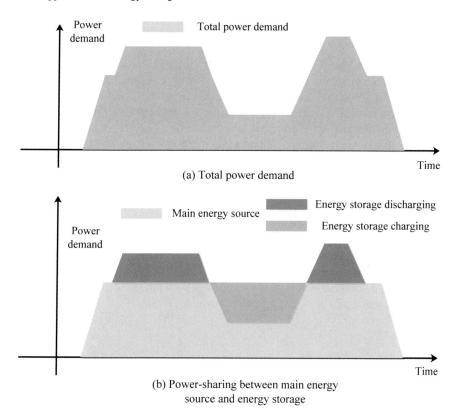

Fig. 6.8 Power sharing by energy storage

the economic and environmental behaviors may be both improved. In this sense, energy storage integration has been viewed as an important approach to facilitate the operation of maritime grids.

It should be noted that energy storage can level/smooth other types of power demand in maritime grids as well, such as service load [3], or weapon system [26], and even in some short-term timescale applications [5–7]. In those applications, the effects are similar to Fig. 6.8a, b, i.e., the main energy source keeps a nearly constant power output and the energy storage shares the fluctuated load demand by continuous discharging/charging. This advantage also gives a new requirement for energy storage management, i.e., the energy storage should coordinate with the main energy source to achieve economic and environmental tasks.

6.3.3 Renewable Energy Integration

To resolve the bottleneck of energy efficiency problems in maritime grids, renewable energy has been gradually integrated into and may soon become an essential part of maritime grids. However, as we have mentioned in Chaps. 1 and 4, the renewable energy is less controllable compared with conventional energy, and the power outputs are generally fluctuating all the time and cannot be accurately forecasted. There are many routes to mitigate the influences of renewable energy and energy storage integration is an important way [24, 27, 28]. Reference [24] gives a schematic diagram of battery energy storage to mitigate the wind power fluctuations, which is shown in Fig. 6.9.

From Fig. 6.9, the battery units are installed with the wind turbine in parallel. Two layers of control strategy are used to determine the battery power for compensating the wind power fluctuation. In the first layer, the wind power is measured and the fluctuation mitigation control layer determines the compensating power. Then the power allocation control layer split the power into each battery unit, including the charging/discharging states and power values. With this compensation, the power output fluctuation of a wind turbine can be greatly reduced.

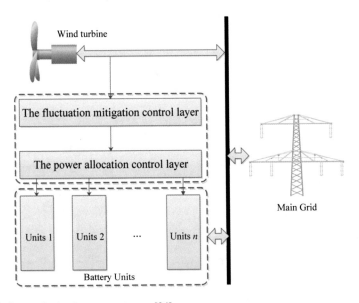

Fig. 6.9 Power sharing by energy storage [24]

6.3.4 Energy Recovery for Equipment

With the electrification of various equipment in maritime grids, energy storage can be used as an energy buffer to recover the wasted energy for later usage. Binti Ahamad et al. [13] has studied the energy recovery by energy storage for an electrified port crane. Figure 6.10 shows 8 working steps for an electrified port crane. The corresponding power demand is shown in Fig. 6.11.

A typical working process of a port crane includes (1) hoist, or beginning to lift up; (2) lifting up speedily; (3) lifting up speedily and the trolley moving forward; (4) lifting up with the full speed and the trolley moving forward; (5) lifting up with slowing speed and the trolley moving with full speed; (6) the trolley moving with slowing speed; (7) lifting down speedily and the trolley moving with slowing speed; (8) settling down. Step (2) and (3) usually have the biggest power demand whereas steps (6), (7) and (8) have smaller power demands. Furthermore, when the cargo is lifting down, the gravitational potential of cargo is wasted, which accounts for about 20% of the total energy consumption [13].

Reference [13] uses a flywheel to store the energy when the cargo is lifting down. The entire process consists of three modes, including mode 1: grid provides power and flywheel discharge; mode 2: grid provides power and flywheel charges; and mode 3: crane charges the flywheel, and three modes are shown in Fig. 6.12. The fourth sub-figure shows the operating cycle of the flywheel.

In Fig. 6.12, mode 1 is used when the power demand is high, and mode 2 is used when the power demand is low, and mode 3 is used when the cargo is lifting down.

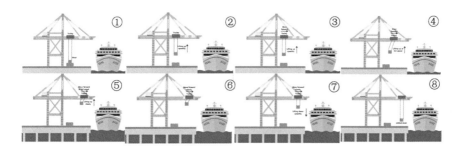

Fig. 6.10 Typical working steps for a port crane

Fig. 6.11 Power demand curves for a port crane. Reprinted from [29], open access

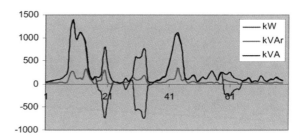

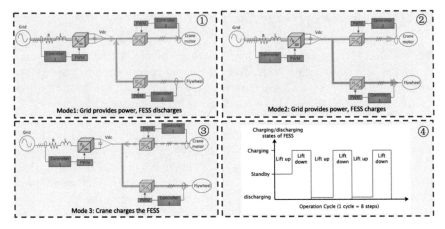

Fig. 6.12 Power demand curves for a port crane [13]

From the overall scope, the flywheel has a periodical operation pattern between "discharging-charging-standby" to recover energy. In a seaport microgrid, there will be increasing electrified equipment and many of them are used for the lifting up/lifting down cargos. Therefore energy storage will be widely used in the future.

6.4 Typical Problems

6.4.1 Energy Storage Management in AES for Navigation Uncertainties

(1) Voyage scheduling and navigation uncertainties

In general, the navigation uncertainty forecasting includes pre-voyage forecasting and intra-voyage forecasting [30]. Responding to the pre-voyage forecasting navigation uncertainties is widely known as the weather routing problems, or pre-voyage planning [30–32]. But the conventional ships are rather difficult to respond to the intra-voyage navigation uncertainties, since in conventional ships, the prime motors are connected with propellers via shafts and gearboxes, and the speed regulation ability of conventional ships are therefore limited. With the development of electric propellers, the prime motors can be "physically separated" from the propellers by the shipboard electric network. With the aid from integrated ESSs, the onboard generation of AESs can quickly and economically respond to the intra-voyage navigation uncertainties. In the future AESs, both the pre-voyage and intra-voyage navigation uncertainties should be addressed by proper energy management.

(2) Two-stage scheduling framework

As shown in Fig. 6.13, the first stage is to respond to the pre-voyage navigation uncertainties and gives the on/off states of onboard DGs, and the second stage is to respond to the intra-voyage navigation uncertainties and gives the loading factors of onboard DGs and other decision variables. The merits are as follow:

a. The two-stage operation model can respond to the pre-voyage navigation uncertainties and intra-voyage navigation uncertainties, coordinately, to gain a compromise between the robustness and flexibility, i.e., the first stage for the worst operating case (robustness) and the second stage to adapt to the current operating case (flexibility).

b. With the proposed two-stage operation, the management of onboard DGs can be more convenient, since the on/off states of onboard DGs are determined before a voyage. The arrangements of the repair or overhaul of the onboard DGs are much easier.

In the pre-voyage time-window, i.e., the first stage, the decision variables are optimized based on the pre-voyage forecasting navigation uncertainty set. The decision variables in the first stage include on/off states of onboard DGs and their loading factors, the shipboard ESS power, the propulsion load and the cruising speed. This stage is to find an optimal robust shipboard operating scheme for addressing the worst speed loss case caused by navigation uncertainties. In this stage, only the on/off states of DGs are "here-and-now" variables and remain as constants in the second stage. Other variables, including the loading factors of DGs, shipboard ESS power, propulsion load, and cruising speed are all "wait-and-see" variables, which will be re-dispatched in the second stage towards uncertainty realization. In the intra-voyage time-window, i.e., the second stage, the navigation uncertainties are treated as realized. All of the "wait-and-see" variables are re-dispatched to address the short-term navigation uncertainties.

The proposed two-stage robust model can be viewed as a "predictive-corrective" process. The first stage is the predictive process to respond to the worst-case and the second stage is the corrective process which takes recourse actions to compensate for the first stage, i.e., reducing the conservatism of the first stage.

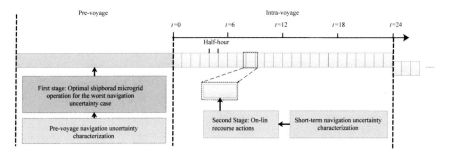

Fig. 6.13 Relation between the first and second stage of proposed model, reprinted from [33], with permission from IEEE

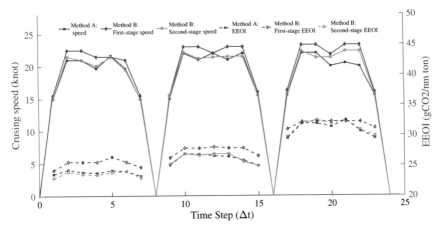

Fig. 6.14 Cruising speed and EEOI comparisons. Reprinted from [33], with permission from IEEE

(3) Case study

To test the proposed two-stage robust optimization problems. Two methods are compared as follows, and the cruising speed and EEOI comparisons are shown in Fig. 6.14.

Method A (Non-robust model): shipboard generation scheduling with the expected wave and wind.

Method B (Robust model): the proposed robust shipboard generation scheduling (first stage and second stage models). In the second stage, an uncertainty sample is selected from the uncertainty set to represent the uncertainty realization.

Firstly, the on-time rates are obtained by generating 500 navigation uncertainty samples in the uncertainty set. The voyage distance of each sample in the terminal port is shown in Fig. 6.15.

In the proposed two-stage robust model, the cruising speed will increase compared with the non-robust model since the robust model is to meet the worst case of the navigation uncertainties, meanwhile, the non-robust model only needs to cope with the expected uncertainties. In this sense, the non-robust model cannot guarantee the on-time rates of AES.

To analyze the effects of energy storage on the navigation uncertainties, the total battery power and SOC in the first and second stages are shown in Fig. 6.16.

From Fig. 6.16, since the worst-case assumed in the first stage may not happen, the total battery power is reduced in the second stage. This phenomenon also shows that the proposed two-stage model can well adapt to the uncertainties with sufficient flexibility.

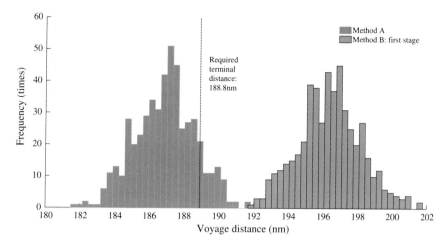

Fig. 6.15 On-time rates of robust and non-robust models. Reprinted from [33], with permission from IEEE

Fig. 6.16 Multi-battery ESS scheme in first and second stages. Reprinted from [33], with permission from IEEE

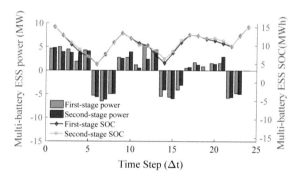

6.4.2 Energy Storage Management in AES for Extending Lifetime

(1) Definitions of DoD and MSOC

In general, improper cycling conditions are the main reasons for battery degradation, i.e., charging/discharging cycles and the DoD in each cycle [34–37]. In recent years, the impacts of MSOC on the battery lifetime have been gradually realized, but still not been incorporated into the operation problem, yet. In fact, DoD and MSOC are two main factors we considered in the battery degradation. The DoDs and initial/terminal SOCs of battery in discharging/charging events are defined in Fig. 6.17a, b.

In Fig. 6.17, when a charging/discharging event begins, the SOC of battery is denoted as the initial SOC, and when this event terminates, the SOC is denoted as

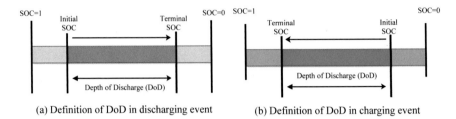

(a) Definition of DoD in discharging event   (b) Definition of DoD in charging event

Fig. 6.17 Definitions of the DoD and initial/terminal SOCs. Reprinted from [38], with permission from IEEE

the terminal SOC. The SOC variations between the initial and terminal SOCs are defined as the DoD, denoted as d. The middle point of the initial and terminal SOCs is defined as the MSOC, denoted as SOC^{mean}.

Since the ship generally has multiple batteries, for the b-th battery in the i-th charging/discharging event, the DoD is denoted as d_i^b, and the corresponding MSOC is denoted as $SOC_{b,i}^{mean}$, and the equivalent life cycle (ELC) is denoted as $ELC_{b,i} = \sum_i d_i^b$.

In the following, we use a vector to denote the MSOC-DoD combination hereafter, i.e., $(SOC_{b,i}^{mean}, d_i^b)$. For example, $(0.3, 0.6)$ means the experiment is conducted in $SOC_{b,i}^{mean} = 0.3$ and $d_i^b = 0.6$.

(2) Impacts of DoD and MSOC on the battery lifetime

In the former section, two main factors for battery degradation have been defined, i.e., d_i^b and $SOC_{b,i}^{mean}$. In the following, a battery degradation model is formulated based on the above two factors. The original dataset is based on experimental research of battery health [39]. It has 14 aging experiments for the batteries in the same brand. The discharging/charging current in each experiment is the same and there are five MSOC-DoD combinations, i.e., $(0.3, 0.6)$, $(0.5, 0.2)$, $(0.5, 0.6)$, $(0.5,1)$, $(0.7, 0.6)$. Several experimental data are shown in Fig. 6.18a, b. If the MSOC-DoD combination is the same, it refers to the experiment that has been conducted twice, otherwise, the experiment is only conducted for once.

In Fig. 6.18, the horizontal axis represents the ELC. The vertical axis represents the normalized battery capacity, and it will decay with the charging/discharging cycles. From above, the impacts of DoD and MSOC on the battery lifetime are clear, i.e., smaller DoD and lower MSOC lead to smaller battery degradation. The reasons are shown as follows, (1) in Fig. 6.18a, experiment 1–6 share the same MSOCs but the DoDs are different, i.e., from 0.2 to 1. As shown in the dataset, the battery with higher DoD will have faster degradation, and (2) in Fig. 6.18b, experiment 7–10 share the same DoD but the MSOC are different, from 0.3 to 0.7. As shown in the dataset, the battery with higher MSOC suffers from higher battery degradation.

To show an example for battery degradation calculation, we take the curve of experiment 6 as an example. The battery in experiment 6 has 879 cycles before life ending. Then the average degradation in each cycle in p.u. is $De_i^b = \frac{1-0.8}{879} =$

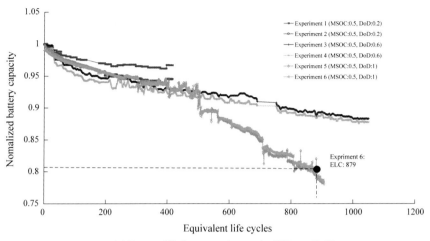

(a) Battery lifetime experiments in different DoDs

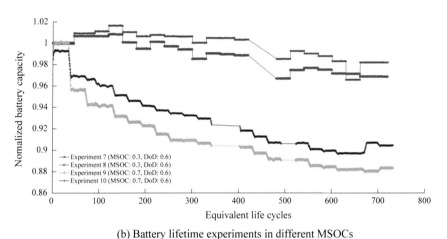

(b) Battery lifetime experiments in different MSOCs

Fig. 6.18 Experimental data illustration. Reprinted from [38], with permission from IEEE

2.2×10^{-4}. Similarly, the average degradations in each cycle for 14 experiments are calculated. For clarification, we denote the obtained battery degradation dataset as $\left(SOC_{b,i}^{mean}, d_i^b, De_i^b\right), b \in \mathcal{B}$, where De_i^b is the average battery degradation.

(3) A revised data-driven battery degradation model

According to Ref. [34], the model of battery lifetime versus DoD is shown as (6.1), where k_1, k_2 and k_3 are all fitting coefficients. To reflect the impacts of MSOC, the degradation model shown in Eq. (6.1) should be revised, and Table 6.3 shows different fitting models and their R-square parameters under the dataset [39]. The fitting tools used is the "sftool" in Matlab 2016b.

Table 6.3 Different Fitting Models and parameters

Model	Model formulation	Fitting parameters	R-square
1	$k_1 \cdot SOC_{b,i}^{mean} \cdot \left(d_i^b\right)^{k_2} \cdot e^{k_3 \cdot d_i^b}$	$k_1, k_2, k_3 = 1.475, -1.106, 3.512$	0.91
2	$k_1 \cdot \left(SOC_{b,i}^{mean}\right)^{k_2} \cdot \left(d_i^b\right)^{k_3} \cdot e^{k_4 \cdot d_i^b}$	$k_1, k_2, k_3, k_4 = 1.41, 1.1, -1.16, 3.62$	0.91
3	$k_1 \cdot e^{k_2 \cdot SOC_{b,i}^{mean}} \cdot \left(d_i^b\right)^{k_3} \cdot e^{k_4 \cdot d_i^b}$	$k_1, k_2, k_3, k_4 = 0.18, 2.0, -1.29, 3.89$	0.87

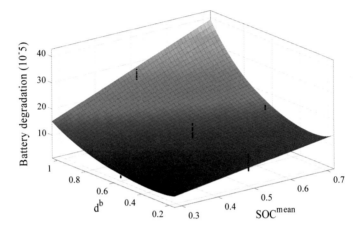

Fig. 6.19 Fitting surface of Battery degradation vs. DoD and MSOC. Reprinted from [38], with permission from IEEE

$$De_i^b = k_1 \cdot \left(d_i^b\right)^{k_2} \cdot e^{k_3 \cdot d_i^b} \tag{6.1}$$

From the results of Table 6.3, model 1and 2 share the best R-square 0.91 with its maximum equal 1, and model 1 is selected as the final battery degradation model since fewer fitting variables and shown as the following Fig. 6.19.

In Fig. 6.19, the black points are the original dataset points, and the fitting surface has shown clear dependence of DoD and MSOC on battery degradation, i.e., higher MSOC and larger DoD will cause higher battery degradation. With the above battery degradation model, the lifetime of battery can be shown as (6.2).

$$L_i^T = \frac{1 - 0.8}{De_i^b} = \frac{0.2}{De_i^b} \tag{6.2}$$

where $1 - 0.8$ means the battery lifetime terminates when the normalized battery capacity becomes 0.8 of its full capacity; L_i^T is the battery lifetime under charging/discharging event i. Obviously, if we want to extend the battery lifetime L_T, De_i^b should be minimized.

(4) Multi-battery scheduling

For indicating when and how many batteries should be utilized, a task matrix B^A is defined and Eq. (6.3) gives an example with the entire operating period having 4 time-intervals, i.e., $t_1 \sim t_4$, and the shipboard ESS include 4 batteries, i.e., no. 1–4.

$$B^A = \begin{bmatrix} 1 & 1 & 0 & 0 \\ 1 & 1 & 0 & 0 \\ 0 & 1 & 1 & 1 \\ 0 & 1 & 1 & 1 \end{bmatrix} \tag{6.3}$$

In (6.3), the row represents batteries and the column represents time-intervals. $B^A(i, j) = B_{i,j}^A = 1$ represents battery i will be switched-on to share power demand (charging or discharging) in the j-th time-interval, or $B^A(i, j) = B_{i,j}^A = 0$ represents the battery i will stand by. With the above definition, the process of multi-battery management can be shown in Fig. 6.20.

With the above multi-battery ESS management, different batteries or battery groups can share different charging/discharging events, which has the potential to reduce the cycles of each battery. The overall lifetime of multi-battery ESS maybe therefore extended.

(5) Case study

To show the benefits of the proposed model, three methods are compared with each other.

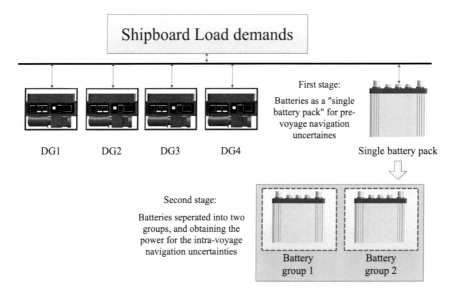

Fig. 6.20 Multi-battery management into two-stage operation. Reprinted from [38], with permission from IEEE

Method A: Conventional energy management without considering battery lifetime degradation [2].

Method B: Conventional energy management only considering DoD as the battery lifetime determinant [34].

Method C: Proposed energy management without multi-battery management.

It should be noted that methods A–C are used to calculate the battery power, and the battery degradations of three methods are calculated by the same model proposed (Model 1 in Table 6.3). The power and SOC curves of three methods are shown in Fig. 6.21, and the corresponding battery degradation in each voyage and overall lifetime are shown in Table 6.4.

In this case study, the voyage is divided into two parts, i.e., $t = 0 \sim 35$ and $t = 41 \sim 64$ are in cruising states, and $t = 36 \sim 40$ is in berthed-in state. From the results in Fig. 6.21, batteries in method A–C all tend to discharge when cruising states to share the power demand and to charge when berthed-in state. It is mainly because when berthed-in, the propulsion load is zero and to avoid frequent start-ups/shut-downs of onboard generators, the energy will be stored in the battery for later usage.

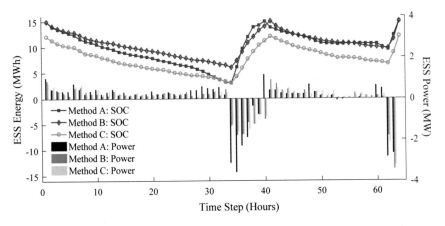

Fig. 6.21 Battery power schedules and SOCs of three cases. Reprinted from [38], with permission from IEEE

Table 6.4 Battery degradation and lifetime in three methods, reprinted from [38], with permission from IEEE	Method	Battery degradation ($\times 10^{-5}$)	Battery lifetime (Times)	Lifetime extension ratios (%)
	A	34.94	572.4	–
	B	25.76	776.4	35.6%
	C	23.0	869.5	51.9%

However, with different battery degradation model, method A–C have different DoDs and MSOCs. In method A, the battery degradation is not considered, the battery operating scheme tends to fully use the battery to reduce $FC^{DG} + FC^{ST}$, and the DoDs of $t = 0$–35, $t = 36$–40 and $t = 41$–64 are 0.8, 0.79, and 0.27, respectively. Meanwhile, in method B, the DoD is considered as the only decision variable of battery degradation. Then the battery operating scheme tends to limit the DoDs, in which the DoDs of $t = 0$–35, $t = 36$–40, $t = 41$–64 decrease to 0.6, 0.6, and 0.27. As a result, the battery lifetime of method B increases by 35.6% compared with method A from Table 6.4. This phenomenon clearly shows that DoD is an important factor for battery lifetime.

In the proposed model (method C), the DoD and MSOC are considered as two factors for battery lifetime. Then compared with method B, method C reduces the MSOC of $t = 0 \sim 35, t = 36 \sim 40$, and $t = 41 \sim 64$ from 0.8, 0.8, and 0.87 (method B) to 0.49, 0.5, and 0.66 (method C). Correspondingly, the battery lifetime of method C increases by 51.9% compared with method A, and 12% longer than method B.

The above phenomenon clearly shows that both the DoD and MSOC have vital impacts on the battery lifetime. The proposed test case has 4 batteries, and each battery has 4 MWh capacity and 2.5 MW power, which is denoted as 1–4. 4 batteries are in two groups. Battery 1, 2 are group 1, and battery 3, 4 are group 2. Method D is designed to show the advantages of multi-battery management. The battery power of methods C and D are shown in Fig. 6.22.

Method D: Proposed energy management with multi-battery management.

From Fig. 6.22, with the multi-battery management, the power demand in different time periods is shared by battery $1 + 2$ and $3 + 4$, respectively. For example, when $t = 0 \sim 13$, the power demand is undertaken by battery $1 + 2$, and when $t = 14 \sim 36$, battery $3 + 4$ undertake the power demand. With this strategy, the battery degradations are shown in Table 6.5.

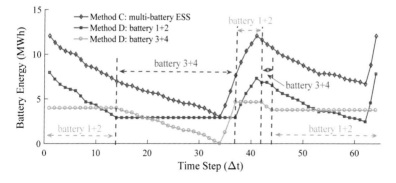

Fig. 6.22 Battery power schedules and SOCs of method C and D. Reprinted from [38], with permission from IEEE

Table 6.5 Battery degradation and lifetime in method C and D, reprinted from [38], with permission from IEEE

Method		Battery degradation ($\times 10^{-5}$/MWh)	Actual battery degradation (MWh)	Battery lifetime (times)
C	1 + 2	11.5	1.84×10^{-3}	869.5
	3 + 4	11.5	1.84×10^{-3}	
	total	23.0	3.68×10^{-3}	
D	1 + 2	27.4	2.19×10^{-3}	980.3
	3 + 4	13.4	1.07×10^{-3}	
	total	20.4	3.26×10^{-3}	

From the above results, the implementation of multi-battery management can further reduce the MSOC of battery 3 + 4, which leads the battery 3 + 4 only have 1.07×10^{-3} MWh degradation compared with 1.84×10^{-3} in method C. As a result, battery 1 + 2 must undertake more power demand than battery 3 + 4, so their degradations increase to 2.19×10^{-3}. In total, the battery degradation in method D is still lower than method C.

In the next voyage, battery 1 + 2 and battery 3 + 4 will change their roles. Battery 1 + 2 will lower their MSOC and battery 3 + 4 will undertake more power to protect the health of battery 1 + 2. With this strategy, the multi-battery management can further extend the total battery lifetime by 12.7%, and the lifetime increases from 869.5 cycles to 980.3 cycles.

As above, the proposed shipboard multi-battery management method can be viewed as a coordinated operation of all the onboard batteries. One battery group undertakes most of the power demand and make the other one working in an MSOC with lower degradation. Then in the next voyage, the battery groups change their roles for the iterative usages.

References

1. Zhao, H., Wu, Q., Hu, S., et al.: Review of energy storage system for wind power integration support. Appl. Energy **137**, 545–553 (2015)
2. Kanellos, F.D.: Optimal power management with GHG emissions limitation in all-electric ship power systems comprising energy storage systems. IEEE Trans. Power Syst. **29**(1), 330–339 (2013)
3. Kanellos, F.D., Tsekouras, G.J., Hatziargyriou, N.D.: Optimal demand-side management and power generation scheduling in an all-electric ship. IEEE Trans. Sustain. Energy **5**(4), 1166–1175 (2014)
4. Fang, S., Xu, Y., Li, Z., et al.: Optimal sizing of shipboard carbon capture system for maritime greenhouse emission control. IEEE Trans. Ind. Appl. **55**(6), 5543–5553 (2019)
5. Hou, J., Sun, J., Hofmann, H.: Mitigating power fluctuations in electric ship propulsion with hybrid energy storage system: design and analysis. IEEE J. Oceanic Eng. **43**(1), 93–107 (2017)

6. Hou, J., Sun, J., Hofmann, H.: Control development and performance evaluation for battery/flywheel hybrid energy storage solutions to mitigate load fluctuations in all-electric ship propulsion systems. Appl. Energy **212**, 919–930 (2018)
7. Hou, J., Song, Z., Park, H., et al.: Implementation and evaluation of real-time model predictive control for load fluctuations mitigation in all-electric ship propulsion systems. Appl. Energy **230**, 62–77 (2018)
8. Kanellos, F.D.: Real-time control based on multi-agent systems for the operation of large ports as prosumer microgrids. IEEE Access **5**, 9439–9452 (2017)
9. Kanellos, F.D., Volanis, E.S., Hatziargyriou, N.D.: Power management method for large ports with multi-agent systems. IEEE Trans. Smart Grid **10**(2), 1259–1268 (2017)
10. Gennitsaris, S.G., Kanellos, F.D.: Emission-aware and cost-effective distributed demand response system for extensively electrified large ports. IEEE Trans. Power Syst. **34**(6), 4341–4351 (2019)
11. Molavi, A., Shi, J., Wu, Y., et al.: Enabling smart ports through the integration of microgrids: a two-stage stochastic programming approach. Appl. Energy **258**, 114022 (2020)
12. Zhao, N., Schofield, N., Niu, W.: Energy storage system for a port crane hybrid power-train. IEEE Trans. Transp. Electrif. **2**(4), 480–492 (2016)
13. Binti Ahamad, N., Su, C., Zhaoxia, X. et al.: Energy harvesting from harbor cranes with flywheel energy storage systems. IEEE Trans. Industry Appl. **55**(4), 3354–3364
14. Díaz-González, F., Sumper, A., Gomis-Bellmunt, O., et al.: A review of energy storage technologies for wind power applications. Renew. Sustain. Energy Rev. **16**(4), 2154–2171 (2012)
15. Mutarraf, M., Terriche, Y., Niazi, K., et al.: Energy storage systems for shipboard microgrids-a review. Energies **11**(12), 3492 (2018)
16. Park, J., Kalev, C., Hofmann, H.: Control of high-speed solid-rotor synchronous reluctance motor/generator for flywheel-based uninterruptible power supplies. IEEE Trans. Industr. Electron. **55**, 3038–3046 (2008)
17. Cheng, M., Sami, S., Wu, J.: Benefits of using virtual energy storage system for power system frequency response. Appl. Energy **194**, 376–385 (2017)
18. Díaz-González, F., Sumper, A., Gomis-Bellmunt, O., et al.: Energy management of flywheel-based energy storage device for wind power smoothing. Appl. Energy **110**, 207–219 (2013)
19. Spiryagin, M., Wolfs, P., Szanto, F., et al.: Application of flywheel energy storage for heavy haul locomotives. Appl. Energy **157**, 607–618 (2015)
20. Norway electric ferry cuts emissions by 95%, costs by 80%. https://reneweconomy.com.au/norway-electric-ferry-cuts-emissions-95-costs-80-65811/
21. A new all-electric cargo ship with a massive 2.4 MWh battery pack launches in China. https://electrek.co/2017/12/04/all-electric-cargo-ship-battery-china/
22. Deep Ocean-01 official ship launched in Shenzhen. http://www.sz.gov.cn/cn/xxgk/zfxxgj/tpxw/content/post_8012934.html
23. 5 MW energy storage for cold-ironing in Lianyugang port. https://www.energytrend.cn/news/20200116-81127.html
24. Jiang, Q., Gong, Y., Wang, H.: A battery energy storage system dual-layer control strategy for mitigating wind farm fluctuations. IEEE Trans. Power Syst. **28**(3), 3263–3273 (2013)
25. Fang, S., Xu, Y., Li, Z., et al.: Two-step multi-objective management of hybrid energy storage system in all-electric ship microgrids. IEEE Trans. Veh. Technol. **68**(4), 3361–3373 (2019)
26. Alafnan, H., Zhang, M., Yuan, W., et al.: Stability improvement of DC power systems in an all-electric ship using hybrid SMES/battery. IEEE Trans. Appl. Supercond. **28**(3), 1–6 (2018)
27. Yao, J., Li, H., Liao, Y., et al.: An improved control strategy of limiting the DC-link voltage fluctuation for a doubly fed induction wind generator. IEEE Trans. Power Electron. **23**(3), 1205–1213 (2008)
28. Li, X., Hui, D., Lai, X.: Battery energy storage station (BESS)-based smoothing control of photovoltaic (PV) and wind power generation fluctuations. IEEE Transactions on Sustainable Energy **4**(2), 464–473 (2013)

29. Tran, T.K.: Study of electrical usage and demand at the container terminal. Deakin University (2012)
30. Perera, L., Soares, C.: Weather routing and safe ship handling in the future of shipping. Ocean Eng. **130**, 684–695 (2017)
31. Krata, P., Szlapczynska, J.: Ship weather routing optimization with dynamic constraints based on reliable synchronous roll prediction. Ocean Eng. **150**, 124–137 (2018)
32. Grifoll, M., Martorell, L., de Osés, F.: Ship weather routing using pathfinding algorithms: the case of Barcelona-Palma de Mallorca. Trans. Res. Procedia **33**, 299–306 (2018)
33. Fang, S., Xu, Y., Wang, H. et al.: Robust operation of shipboard Microgrids with multiple-battery energy storage system under navigation uncertainties. IEEE Trans. Vehic. Technol., In Press (2020)
34. Ju, C., Wang, P., Goel, L., et al.: A two-layer energy management system for microgrids with hybrid energy storage considering degradation costs. IEEE Trans. Smart Grid **9**(6), 6047–6057 (2017)
35. Zhou, C., Qian, K., Allan, M., et al.: Modeling of the cost of EV battery wear due to V2G application in power systems. IEEE Trans. Energy Convers. **26**(4), 1041–1050 (2011)
36. Farzin, H., Fotuhi-Firuzabad, M., Moeini-Aghtaie, M.: A practical scheme to involve degradation cost of lithium-ion batteries in vehicle-to-grid applications. IEEE Trans. Sustain. Energy **7**(4), 1730–1738 (2016)
37. Liu, K., Li, Y., Hu, X., et al.: Gaussian process regression with automatic relevance determination kernel for calendar aging prediction of lithium-ion batteries. IEEE Trans. Industr. Inf. **16**(6), 3767–3777 (2019)
38. Fang, S., Gou, B., Wang, Y., et al.: Optimal hierarchical management of shipboard multi-battery energy storage system using a data-driven degradation model. IEEE Transactions on Transportation Electrification **5**(4), 1306–1318 (2019)
39. Advanced Life Cycle Engineering (CALCE) at the University of Maryland. https://web.calce.umd.edu/batteries/data.htm

Chapter 7
Multi-energy Management of Maritime Grids

7.1 Concept of Multi-energy Management

7.1.1 Motivation and Background

Generally, all the energy systems are "multi-energy systems" in the sense that multiple energy sectors interact at different levels. For example in conventional power systems, the coal or gas used for generating electricity should be transported to each power plant, and this process implies the couplings between fossil energy and electrical energy. Another case is, the heating service by the combined heat-power plant also last for decades, and this process includes the coupling between heating energy and electrical energy. However, those energy couplings between different systems are conventionally weak compared with the relationship within a single energy system, and that is the main reason for the past studies of power system mostly only consider the electrical energy [1–3]. However, the interactions between different energy systems become tighter and more frequent recently, and this trend is about to continue in the future [4–7], such as the electric-gas energy system, and the coordinated heat-power system, or even the transportation-power system motivated by the transportation electrification. In this sense, conventional energy management for a single energy system may not be valid in the future, which drives the research of multi-energy management.

In literature, [8–11] focus on the coordination between the gas system and power systems [12–15]. Study the energy management methods for heat-power systems [16, 17]. Study the water-power systems and [18–22] investigate the coupling between the transportation system and power system by electric vehicles' charging and discharging. The above research has brought a new perspective in energy system analysis, particularly in the light of reducing the economic and environmental burden of energy services. In summary, three benefits can be achieved by multi-energy management:

S. Fang and H. Wang, *Optimization-Based Energy Management for Multi-energy Maritime Grids*, Springer Series on Naval Architecture, Marine Engineering, Shipbuilding and Shipping 11, https://doi.org/10.1007/978-981-33-6734-0_7

a. Increasing or improving the energy efficiency of the entire system and the utilization of primary energy sources. The reason is the multi-energy system can use the energy at different levels. For example, the waste heat after generating electricity can be used for heating services and the energy efficiency of the entire system improves.

b. Better deploying various energy resources at multiple system levels. For example, small-scale gas turbines can respond to volatile electricity market prices in a wind-rich energy system.

c. Increasing the energy system flexibility by the coordinations between different energy systems. For example, scheduled charging/discharging of the electric vehicles acts as demand response tool for power system. Or the thermal storage tank can bring flexibility for combined power-heat plants.

Since the above three main advantages, the research on multi-energy management is essential for future energy systems. However, different energy systems have different administrators and quite distinct characteristics, and their coordinations are much more complex compared with the coordinations within a single energy system. Proper modeling methods and control strategies should be proposed to facilitate their operation.

7.1.2   Classification of Multi-energy Systems

The multi-energy systems can be classified by different perspectives, and there are mainly four perspectives to characterize the MES. The first is the spatial perspective. This perspective points out how MES can intend at different levels of aggregation in terms of components or even just conceptually. These levels go from buildings to district and finally to regions and even countries. This classification is shown in Fig. 7.1a.

The second perspective focuses on the provision of multiple services by optimally scheduling different energy systems, particularly at the supply levels. Such as the services provided by the MES, including electricity supply, water supply, heating service, EV charging services, gas filling services, and so on. This classification is shown in Fig. 7.1b.

The third perspective highlights how different types of fuels can be integrated together for providing optimal energy services, typically for economic or environmental targets. The fuel types range from classical fossil fuel, such as oil, coal and natural gas, to biomass fuels, and renewable energy. This classification is shown in Fig. 7.1c.

The fourth perspective discusses the coordinations between different energy systems, especially the coordination between different networks, such as the electrical network, gas network, district heating/cooling network, in terms of facilitating the development of multi-energy management methods and their interactions. This classification is shown in Fig. 7.1d.

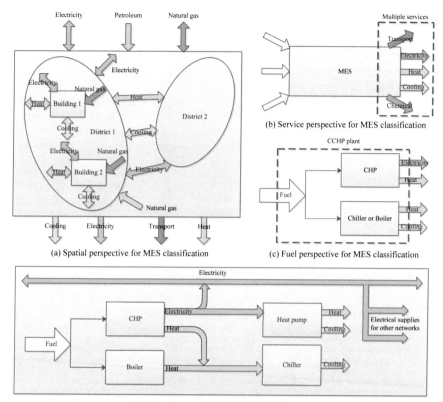

(a) Spatial perspective for MES classification

(b) Service perspective for MES classification

(c) Fuel perspective for MES classification

(d) Network perspective for MES classification

Fig. 7.1 Classifications for MES [4]

Figure 7.1a classifies the MESs from the spatial perspective. An individual building exchanges energy by the transmission of electricity, heat, cooling, and natural gas. Then multiple buildings aggregate as a district, then multiple districts aggregate as a region and expand to a wider area. In this perspective, MESs can be classified as the building MES, district MES, region MES, and so on.

Figure 7.1b classifies the MESs from the service perspective. Generally, MES can provide multiple services to the customers, such as the electricity supply, heat and cooling power, and even some transport services, such as the charging/discharging of EV. In this perspective, MESs can be classified as combined electric-heat MES, combined electric-heat-cooling MES, and even electric-heat-water supply MES, since the water pump is coupled with the electric network by the electrical water pumps.

Figure 7.1c classifies the MES from the fuel perspective. For example, there exist many power sources in MES, such as power plants, boilers, gas turbines, and chillers. They may consume different types of fuels. Different power plants may consume coal, oil, or gas. A boiler may consume electricity or other fossil fuel, and a chiller may consume electricity or heat power. In this sense, the fuel type can also classify the MESs, such as the coal-gas MES, gas-hydrogen MES, or even ammonia MES since ammonia is a new type of carbon-free fuel [23].

Figure 7.1d classifies the MES from the network perspective since every "energy carrier" should be transmitted by a designed network. The electrical network includes power systems on multiple scales. Gas and oil are transported by pipelines or transportation flows. Heat and cooling power also have certain pipelines. Those different networks can have different topologies and operating strategies, which is the main motivation of this classification method. In this sense, the networks of MESs can be classified as combined electric-heat networks, combined electric-heat-cooling networks, and so on.

7.2   Future Multi-energy Maritime Grids

7.2.1   Multi-energy Nature of Maritime Grids

A sketch of MES is given in the former section to show the basic advantages and characteristics. In this section, the multi-energy nature of maritime grids will be analyzed to show their similarities and differences compared with conventional MESs, and Fig. 4.1 is re-drawn below as Fig. 7.2 as an illustration of future maritime grids. Two cases of maritime grids will be given after this illustration.

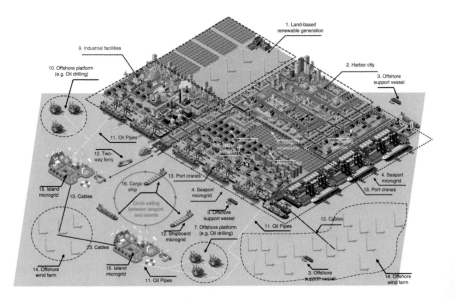

Fig. 7.2   Illustrations of future maritime grids

(1) Spatial perspective

From Fig. 7.2, maritime grids cover different spatial areas. For example, island microgrids cover an individual island, and the energy sources include offshore wind power, photovoltaic power, and underground cables. Seaport microgrids cover the harbor territory, and the energy sources include the offshore wind farm, land-based photovoltaic farm, oil pipelines, and the electricity supply from the harbor city. Other maritime grids include the drilling platforms and different types of ships. In summary, maritime grids have a very wide range on system scales, from the smallest to a ferry or a building and the biggest to a harbor city, which involves all the energy sources within a conventional MES. Different maritime grids are coupled tightly by energy connections, and current multi-microgrid coordination methods can be used in maritime grids to achieve better system characteristics.

(2) Service perspective

Figure 7.2 shows maritime grids can provide different services to customers, including the conventional services of electricity, heat, cooling in land-based MES, also including some types beyond current focuses, such as the logistic services, fuel transportation services. This is the primary difference between current studied MES (land-based MES) from the maritime grids. This is also a challenge for the research of maritime grids, since new energy models of those services should be formulated and integrated into the energy management model.

(3) Fuel perspective

Maritime grids also involve different types of fuels. In Fig. 7.2, the drilling platform can harvest crude oil or natural gas, and transport them to an island or the seaport. The industrial factory can refine crude oil into different types of fuels, such as gasoline, diesel, and so on. Those fuels may in reverse fill into the ships for sailing, into seaport for generation, and into the harbor city for daily lives. Besides, some novel fuels may also use in maritime grids, such as ammonia, methanol, and ethanol.

(4) Network perspective

Maritime grids also have different types of networks. Figure 7.2 shows some typical ones, (1) electrical networks in harbor city, seaport, industrial factory; (2) heat/cooling networks in harbor city, seaport, industrial factory; (3) fossil fuel networks between the ocean platforms and a seaport or an island; (4) electrical networks between offshore wind farms and a seaport or an island; (5) multi-energy network within an island; (6) transportation network by ships and vehicles. Those networks above are connected with multiple energy and information flows and may be more complex than conventional land-based MESs.

7.2.2 Multi-energy Cruise Ships

In Fig. 7.3, a typical topology of a multi-energy cruise ship is shown. Detailed illustrations can be depicted as follow. The load demands can be classified into three categories, the electric load, thermal load, and propulsion load. Among the three load demands, the propulsion load is to drive the cruise ship, which consists most of, usually more than 50% of the total load demand [24]. The propulsion load has a simple cubic relationship with the cruising speed, which is under the constraints of navigation distance [25]. The electric load in cruise ships includes the illumination, recreation equipment, movie theater, and so on. This type of load scales up to tens of megawatt [24], which is provided by the electric power bus, shown as the blue lines and arrows in Fig. 7.3. The thermal load in cruise ship includes the cooling and heat load, the swimming pool, and the cooking. This type of load also scales up to tens of megawatt [27], which is provided by the thermal power network, shown as the green line and arrows in Fig. 7.3. It also should be noted that in some cruise ships the cooling and heat loads are provided by the electricity. In this work, we will compare the introduced multi-energy technology with the single electric supply mode.

As for the generation systems, to provide adequate electric and thermal loads for the overall cruise ship. There exist three types of generation systems, i.e., DG, CCHP, and PTC. The DGs make up the main part of the shipboard generation, which provides most of the electric power supply. The CCHP both provides the electric power and the thermal power and the PTC uses electricity to produce thermal power. To balance both the electric and thermal loads, the HES (electric and thermal energy storage) is integrated.

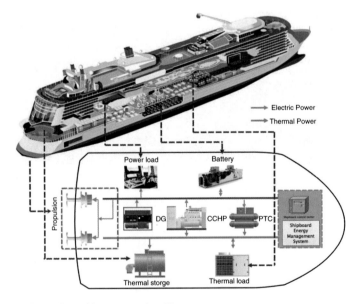

Fig. 7.3 Topology of a multi-energy cruise ship

7.2.3 Multi-energy Seaport

We have illustrated the multi-energy seaport in Chap. 1. Here we re-draw Fig. 1.17 as Fig. 7.4to further show its multiple energy flows.

Generally, the seaport is connected with the main grid and various renewable energy are integrated, i.e., seaport wind farms and PV farms in Fig. 7.4. All the port-side equipment, including the quay cranes, gantry cranes, transferring trunks, are electrically-driven. The seaport provides four types of services to the berthed-in ships and has four sub-systems for each type of services: (1) logistic service. The berth allocation and quay crane scheduling for loading/unloading cargo; (2) fuel transportation. Unloading or refilling fuel for the berthed-in ships; (3) cold-ironing. Providing electricity to the berthed-in ships and (4) refrigeration reefer for the cold-chain supply. The coordination between different sub-systems is shown in Fig. 7.5. Four sub-systems are communicating by the seaport control center and a distributed control strategy is employed in the seaport microgrid.

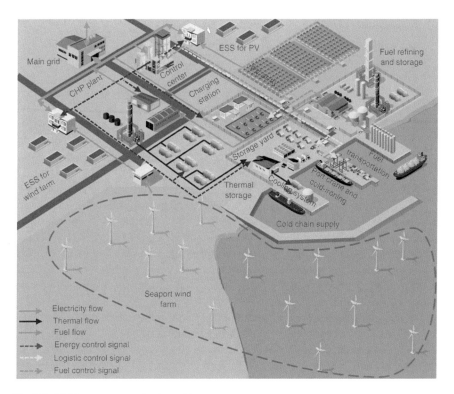

Fig. 7.4 Multi-energy seaport microgrid

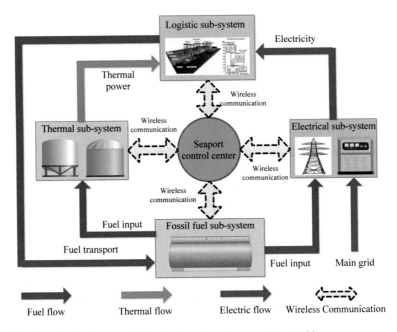

Fig. 7.5   Coordination between different sub-systems in seaport microgrid

7.3   General Model and Solving Method

7.3.1   Compact Form Model

From above, maritime grids involve different networks and provide multiple types of services by different types of fuels. In this sense, maritime grids have a significant characteristic, i.e., using the electric network as the backbone for energy management, and other different networks serve as the "load demand" of electric networks. For example, the heat/cooling networks couple with the electric network by CHP or electric boiler/chiller, and water supply network couple with the electric network by electric water pumps, and logistic network couple with the electric network by charging/discharging.

For this complex network, a general energy scheduling form can be shown as (7.1). Where $f(x)$ is the objective function of the main network, generally the electric network, and x is the decision variable vector; $g_i(y_i)$ is the objective function of the i-th network, and y_i is the decision variable vector of the i-th network; $F(x)$ is the constraint set of the main network; $G_i(y_i)$ is the constraint set of the i-th network; $A_i \cdot x = H_i(y_i)$ is the coupling constraint set of power consumption of coupling equipment, such as water pump, CHP, and various logistic equipment.

$$\min_{x,y_i} f(x) + \sum_{i=1}^{n} g_i(y_i)$$
$$s.t.\, F(x) \geq 0,\, G_i(y_i) \geq 0 \tag{7.1}$$
$$A_i \cdot x = H_i(y_i),\, x \in X,\, y_i \in Y_i$$

7.3.2 A Decomposed Solving Method

This Chapter proposes a decomposed method to solve this type of problem, which is given by the following Theorem 7.1.

Theorem 1 *The above formulation is equivalent to the following form.*

$$\min_{x} \left[f(x) + \sum_{i=1}^{n} \inf_{y_i,\tau_i,u_i} (g_i(y_i) + \tau_i \cdot G_i(y_i) + u_i \cdot [H_i(y_i) - A_i \cdot x]) \right]$$
$$s.t.\, F(x) \geq 0,\, G_i(y_i) \geq 0$$
$$x \in X \cap V \tag{7.2}$$
$$V \equiv \bigcup_{i=1}^{n} \{x | \lambda_i \cdot [H_i(y_i) - A_i \cdot x] = 0\}$$
$$, where \lambda_i \geq 0 \, and \, \sum_{i=1}^{n} \lambda_i = 1$$

where u_i is the optimal multiplier vector of the following optimization problem.

$$\min_{y_i} \sum_{i=1}^{n} g_i(y_i)$$
$$s.t.\, G_i(y_i) \geq 0,\, A_i \cdot x = H_i(y_i),\, for\, all\, i \tag{7.3}$$

Proof

(1) Problem (7.1) and (7.2) have the same feasible region.

(1.1) If $\bar{x}$ be feasible for (7.1), then $\bar{x}$ is feasible for (7.2).

Let $\bar{x}$ be an arbitrary point in the feasible region of (7.1), then

$$F(\bar{x}) \geq 0,\, A_i \cdot \bar{x} = H_i(y_i),\, for \forall i \tag{7.4}$$

It follows that (7.5) holds for all λ_i.

$$\lambda_i \cdot [H_i(y_i) - A_i \cdot \bar{x}] = 0 \tag{7.5}$$

Then $\bar{x} \in V$, and $F(\bar{x}) \geq 0$. $\bar{x}$ is also feasible for (7.2).

(1.2) If $\bar{x}$ be feasible for (7.2), then $\bar{x}$ is feasible for (7.1).

Let $\bar{x}$ be an arbitrary point for (7.2), then (7.5) holds at least for one i. $F(\bar{x}) \geq 0$ is satisfied all the same, then (7.6) holds.

$$\eta \cdot F(\bar{x}) + \lambda_i \cdot [H_i(y_i) - A_i \cdot \bar{x}] \geq 0 \qquad (7.6)$$

It follows that

$$\operatorname*{Inf}_{\eta \geq 0}\{\eta \cdot F(\bar{x}) + \lambda_i \cdot [H_i(y_i) - A_i \cdot \bar{x}] \geq 0\} \qquad (7.7)$$

Since $\eta = 0$ is allowed in (7.7). Now, (7.7) is the dual of the following optimization problem.

$$\begin{aligned} \min_{y_i \in Y_i} 0^T \cdot y_i \\ s.t. F(x) \geq 0, \; H_i(y_i) = A_i \cdot \bar{x} \end{aligned} \qquad (7.8)$$

Obviously, (7.8) is feasible and has the optimal value of 0, hence, $\bar{x}$ is feasible for (7.1).

(2) The objective function
Since u_i is the optimal multiplier vector of (7.3), then (7.9) holds.

$$\min_{y_i} \sum_{i=1}^{n} g_i(y_i)$$
$$= inf\left\{\sum_{i=1}^{n} g_i(y_i) + \sum_{i=1}^{n} \tau_i \cdot G_i(y_i) + \sum_{i=1}^{u} u_i \cdot [H_i(y_i) - A_i \cdot x]\right\} \qquad (7.9)$$

In this sense,

$$\min_{x, y_i}[f(x) + \sum_{i=1}^{n} g_i(y_i)]$$
$$\min_{x}\left[f(x) + \sum_{i=1}^{n} \inf_{y_i, \tau_i, u_i} (g_i(y_i) + \tau_i \cdot G_i(y_i) + u_i \cdot [H_i(y_i) - A_i \cdot x])\right] \qquad (7.10)$$

From above, (7.1) and (7.2) are equivalent, then the solution process is given below.

Solution process: From (7.2), the original problem can be solved in a two-step process. It should be noted that, $g_i(y_i) + \tau_i \cdot G_i(y_i)$ is a constant when minimizing x, so it is eliminated for simplification.

Step 1: Given a feasible $\bar{x}$, solve (7.11) for y_i^* and u_i.

$$\min_{y_i} \sum_{i=1}^{n} g_i(y_i)$$
$$s.t. G_i(y_i) \geq 0, \; A_i \cdot \bar{x} = H_i(y_i), \; for \, all \, i \qquad (7.11)$$

It should be noted that, there are no coupling between different networks. So (7.11) can be solved in parallel.

Step 2: Solve (7.12) for x.

$$\min_x \left[f(x) + \sum_{i=1}^{n} \left(u_i \cdot \left[H_i(y_i^*) - A_i \cdot x \right] \right) \right]$$
$$s.t. F(x) \geq 0, \, A_i \cdot x = H_i(y_i^*), \, for \, all \, i$$
(7.12)

Then check the convergence characteristic, if yes, terminates and if not, return to Step 1 and update $\bar{x}$. The algorithm convergence is given below.

Algorithm convergence It is proved that the proposed method has finite ε-convergence characteristic.

Theorem 2 *Assume X and V are both compact set, f, g, F, G_i and H_i are continuous. The set $UT(x)$ of the optimal multiplier vector for (7.3) is non-empty for all x in X and uniformly bounded. Then, for any given $\varepsilon > 0$, the proposed procedure terminates in a finite number of steps.*

Proof For simplification, we define (7.13).

$$L(x, \tau, u) = f(x) + \sum_{i=1}^{n} (g_i(y_i) + \tau \cdot G_i(y_i) + u \cdot [H_i(y_i) - A_i \cdot x])$$
(7.13)

For any sequence $L(x^v, \tau^v, u^v)$, x^v of the optimal solution of (7.2). Firstly, the optimal multipliers sequence τ^v, u^v will converges to a point noted as $(\bar{\tau}, \bar{u})$, since the uniformly bounded assumption of $UT(x)$. Additionally, x^v will converge to a point denoted as $\bar{x}$ since the compactness of X.

At last, since $L(x^v, \tau^v, u^v)$ is a non-increasing sequence and bounded below, there exists at least one sub-sequence of $L(x^v, \tau^v, u^v)$, x^v which converges to a point, we noted it as $L(\bar{x}, \bar{\tau}, \bar{u})$, $\bar{x}$.

Since the weak duality, $(\bar{\tau}, \bar{u})$ is the optimal multiplier for $\bar{x}$ and (7.14) holds.

$$L(\bar{x}, \bar{\tau}, \bar{u}) = \inf_{y_i} \left(f(\bar{x}) + \sum_{i=1}^{n} g_i(y_i) \right)$$
(7.14)

Then, for any given $\varepsilon > 0$, there should be finite v to make (7.15) hold.

$$L(\bar{x}, \bar{\tau}, \bar{u}) \leq L(x^v, \tau^v, u^v) \leq \inf_{y_i} \left(f(\bar{x}) + \sum_{i=1}^{n} g_i(y_i) \right) + \varepsilon$$
(7.15)

Then the proposed method should converge in finite steps.

7.4 Typical Problems

7.4.1 Multi-energy Management for Cruise Ships

This section uses the cruise ship in Fig. 7.3 as the test case to show the effects of energy management. For a more economic and environmental operation of the cruise ship, the shipboard energy management system will optimally dispatch the outputs of the DG, CCHP, PTC, and HES to fulfill the propulsion, onboard electric, and thermal loads. However, in practice, those control variables are not on the same time-scale. During the navigation, the ship will constantly cruise and the speed cannot be regulated rapidly [25], and the onboard facilities for tourists also should keep working till night. This makes the propulsion and electric loads should be fulfilled in a long-term horizon (every hour in this work). Besides, the thermal load should be satisfied in a short-term horizon (20 min) to meet the real-time constraints of indoor temperature and hot water supply. To coordinately satisfy the above load demands in two time-scales, in this work we propose a two-stage operation framework for the cruise ship, which is shown as follow:

From the Fig. 7.6, the first stage hourly schedules the DGs, CCHP, and battery to fulfill the voyage distance constraints and hourly electric load demand. The thermal power produced by the CCHP is stored in the thermal energy storage. In the second stage, every 20 min, the PTC and thermal energy storage is dispatched to meet the thermal load demand. With the above operation framework, both the propulsion and electric loads can be met in a long-term time-scale, as well as the thermal load demand can be met in a short-term time-scale to improve the QoS.

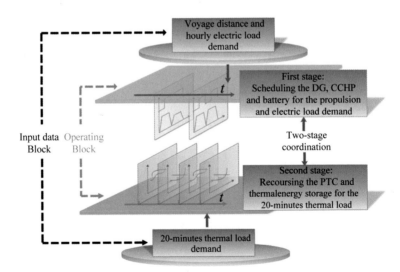

Fig. 7.6 Two-stage operation framework for the cruise ship, reprinted from [26], with permission from IEEE

To show the benefits of the proposed model, the onboard generation and battery SOC are shown in Fig. 7.7a, b, respectively. Figure 7.8 compares the results of the thermal load of the proposed two-stage method.

From Fig. 7.7a, the battery can coordinate with the speed adjustment to smooth the load profiles, which facilitates the economy of cruise ships (the DGs can better operate around their economic points). From Fig. 7.7b, the battery may have much deeper charging/discharging events without the speed adjustment. That is mainly because the cruising speed is fixed during the cruising time-intervals, and the battery should quickly respond to the load profiles for the economy of navigation.

From Fig. 7.8a, the proposed two-stage scheduling model can meet the thermal load demand in a more accurate time-scale by simply dispatching the loading factor

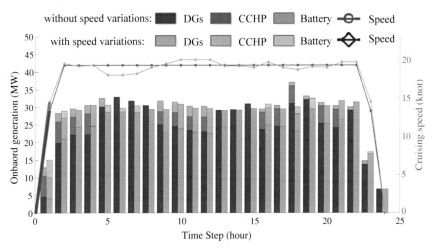

(a) Onboard generation with/without speed variations

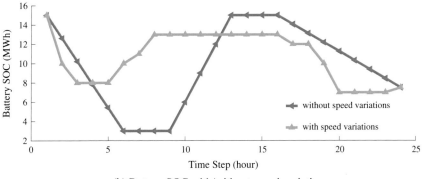

(b) Battery SOC with/without speed variations

Fig. 7.7 Onboard generation and battery SOC with/without speed variations, reprinted from [26], with permission from IEEE

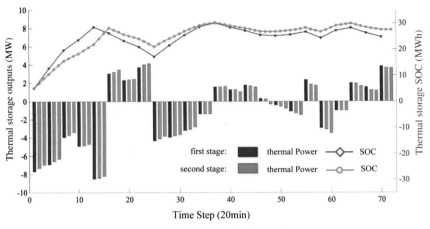

(a) thermal storage in the first/second stages

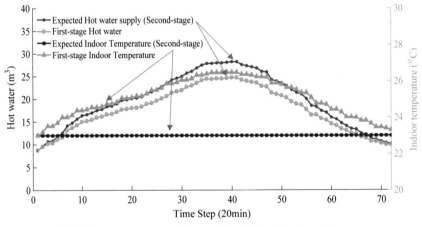

(b) indoor temperature and thermal load in the first/second stages

Fig. 7.8 Onboard thermal storage and thermal load, reprinted from [26], with permission from IEEE

of the thermal storage tank, and the outputs of PTC and CCHP. The results are shown in Fig. 7.8b. The indoor temperature can be kept as a constant meanwhile the single first stage will have a maximal 3 °C temperature variations since the accumulated effects of thermal load demand variations. Similarly, the single first stage also cannot meet the hot water supply-demand all the time, and the thermal variations will also be accumulated and make the supplies always smaller than the demands.

Current cruise ships are mainly BOS cruise ships, which means in the BOS mode, the thermal load demand is all provided by the electric-side (PTC units). In this case,

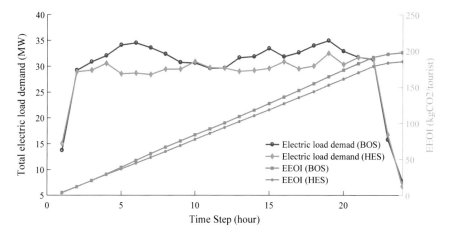

Fig. 7.9 Comparisons between BOS and HES cruise ships, reprinted from [26], with permission from IEEE

the BOS ship replaces the CCHP to conventional DG with the same capacity. The parameters are the same with DG2, 3. The total load demand and EEOI of BOS and HES ships are shown in Fig. 7.9.

From Fig. 7.9, the BOS cruise ship will have much larger load demands since the thermal load is provided by the PTC unit. Correspondingly, the EEOI of the HES integrated cruise ship is also much smaller than the BOS by 8.37%.

7.4.2 Multi-energy Management for Seaport Microgrids

(1) System description

From Fig. 7.10, there are three energy resources in this microgrid, i.e., photovoltaics(PVs), electrical substation, and gas pressure house. The PVs and substation inject electricity into the seaport microgrid via DC and AC buses, respectively. The gas pressure house injects gas into the seaport microgrid to the gas storage. Additionally, to improve the system flexibility, a battery energy storage system (ESS) and two thermal storages are incorporated. The AC/DC loads and heat/cooling power are supplied to the seaport loads, and DC power is used for charging the electric trunk. The power to gas equipment transforms the excess power to gas to fill the gas vehicles.

In this paper, the scheduling horizon is divided into equal time step Δt, denoted by set $\mathcal{T} = \{1, 2, \ldots, T\}$. The proposed operation method is formulated as a two-stage framework, where the first stage is for the day-ahead time-scale, and the second stage is for real-time scheduling, i.e., hourly. In the day-ahead operation (first stage), the hourly energy scheme is provided considering the uncertainties, and then in

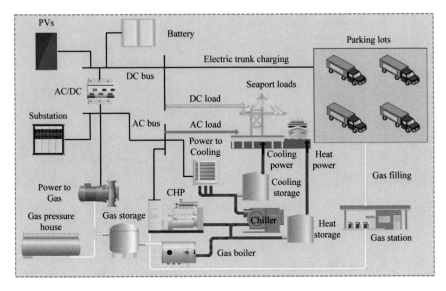

Fig. 7.10 An illustrated seaport microgrid case revised from [28]

the second stage, the seaport microgrid adjusts its scheduling plan responding to the realization of uncertainties in the hourly time-scale. The electrical load profile, heating load profile, and cooling load profile are shown in Fig. 7.11, which are all given in 1000 scenarios. Other detailed parameters can be found in [28].

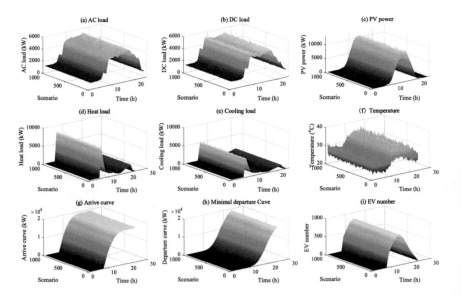

Fig. 7.11 Input parameters of the proposed method, reprinted from [28], open access

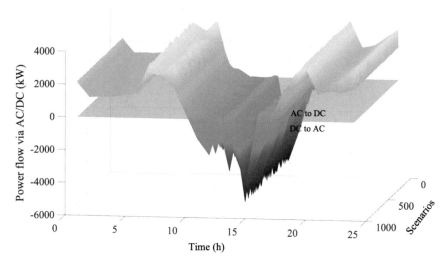

Fig. 7.12 Power flow via bi-directional AC/DC converter, reprinted from [28], open access

(2) Case study

To verify the effectiveness of the proposed method, different cases are formulated as follows.

Case 1: Two-stage optimization is considered, meanwhile the joint constraints are considered.

Case 2: Only the first-stage optimization is considered.

(2.1) Bi-directional AC/DC power flow

To show the coordination between AC and DC sides, the power flow via the bi-directional AC/DC converter is shown in Fig. 7.12. The AC to DC power is shown as the surface above the zero surface, while the DC to AC power is shown as the surface below the zero surface. Then, to show the effects of ESS, the state of charge (SOC) of battery is shown in Fig. 7.13.

From the above figure, at first, when the PV power is almost zero, i.e., $t = 0$–5 h, 20–24 h, the DC load is mainly met by AC to DC converter. When the DC load gradually increases, the AC to DC power is also increasing, and the battery discharges to further support DC load, i.e., $t = 5, 6$ h. After that, with the PV power increasing, the power demands also become larger, i.e., both DC and AC loads during $t = 10$–16 h. In those time intervals, the PV output is beyond the maximal DC load, which leads the PV power change to AC via AC/DC converter to support the AC load or charge into battery, which is shown as the surface below zero in Fig. 7.12 and the charging event in Fig. 7.13. From the above results, the integration of the AC/DC converter can bring great flexibilities to meet both DC and AC loads. The DC power for PV and AC power from UG and CHP can coordinately operate to enhance energy efficiency.

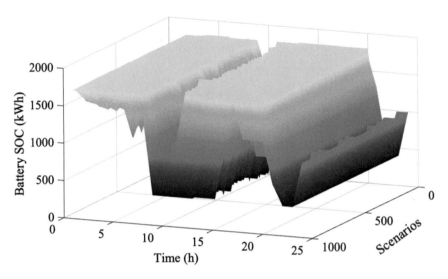

Fig. 7.13 SOC of battery, reprinted from [28], open access

(2.2) Multiple energy flows

In this seaport microgrid, various energy carriers are working coordinately to enhance operation flexibility. To show those coordinations, the power of CHP is shown in Fig. 7.14, the power of heat storage is shown in Fig. 7.15, the power of cooling storage is shown in Fig. 7.16, and the power of power-to-gas facility is shown in Fig. 7.17.

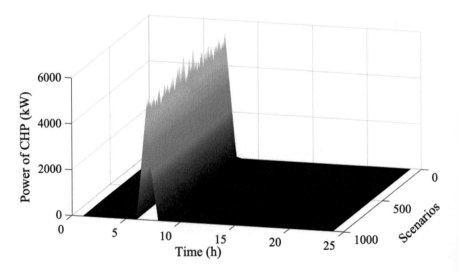

Fig. 7.14 Power of CHP, reprinted from [28], open access

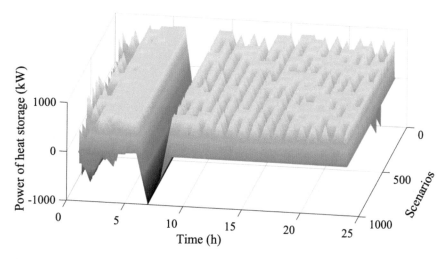

Fig. 7.15 Power of heat storage, reprinted from [28], open access

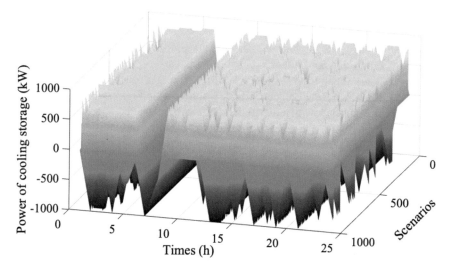

Fig. 7.16 Power of cooling storage, reprinted from [28], open access

From Fig. 7.11d and e, there are two demand impulses of both heat and cooling demands in $t = 6, 7$ h. The CHP responds to those demand impulses and consumes the gas to produce electricity and heat. The heat energy is stored and both the heat and cooling storages are discharging in this period to satisfy the demand, which is shown as the great valleys in their energy curves in Figs. 7.15 and 7.16. After that, CHP is shut-down since the total electricity demand is limited. The thermal demands are then met by the coordination of thermal storage and the gas boiler.

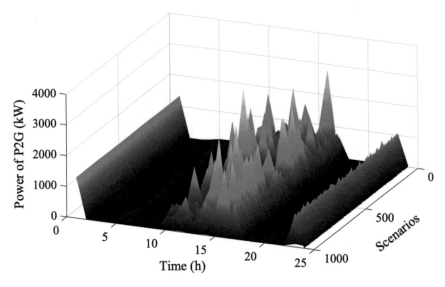

Fig. 7.17 Power of P2G equipment, reprinted from [28], open access

It should be noted that when $t = 10–15$ h, the temperature increases and requires great air-conditioning power demand. While in this time period, the PV power is also in its peak-hours. Then the PV power is converted to gas for the gas boiler to meet the air-conditioning power demand, which is shown as in Fig. 7.17.

The above results show that different energy carriers can be coordinated flexibly in a seaport microgrid. The excess electricity can be converted to gas for thermal demand. With the interactions between different energy carriers, the electric and thermal demand can both be satisfied and the flexibility can be enhanced.

(2.3) Electric and gas trucks

The energy demand of trucks is quite important in future seaport since they play a major role for cargo lifting and transporting. However, before the completed electrification of vehicles, the gas trunks and electric trunks will both exist in seaport microgrid. To satisfy their energy demands, the electric and gas sub-systems of seaport microgrid should be operated in coordination, respectively. In this case, the equivalent energy of gas trucks are shown in Fig. 7.18, and the charging power of electric trucks are shown in Fig. 7.19.

From Fig. 7.18, the energy peaks of gas vehicles are $t = 10–15$ h and 20–24 h. The first peak period corresponds to the working hours, and the second is the vehicles coming back for charging. From the results in Fig. 7.19, the charging patterns are more periodic with three peak hours, i.e., $t = 10–15$, 16–18, and 20–24 h. From the above results, both the gas and electricity demands of trunks can be satisfied.

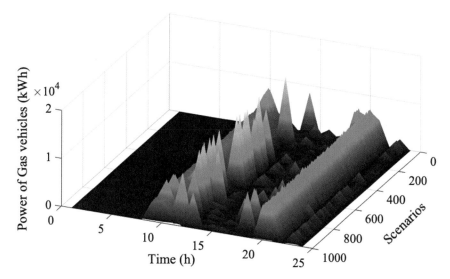

Fig. 7.18 Equivalent energy of gas vehicles, reprinted from [28], open access

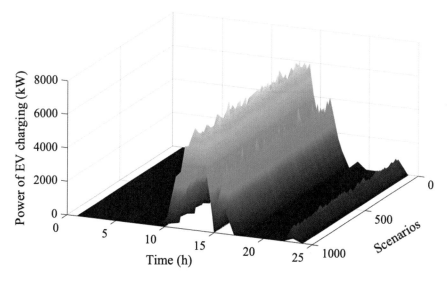

Fig. 7.19 Charging power of electric trunks, reprinted from [28], open access

References

1. Banakar, H., Luo, C., Ooi, B.T.: Impacts of wind power minute-to-minute variations on power system operation. IEEE Trans. Power Syst. **23**(1), 150–160 (2008)
2. Wang, J., Shahidehpour, M., Li, Z.: Security-constrained unit commitment with volatile wind power generation. IEEE Trans. Power Syst. **23**(3), 1319–1327 (2008)
3. Alguacil, N., Motto, A.L., Conejo, A.J.: Transmission expansion planning: A mixed-integer LP approach. IEEE Trans. Power Syst. **18**(3), 1070–1077 (2003)
4. Mancarella, P.: MES (multi-energy systems): an overview of concepts and evaluation models. Energy **65**, 1–17 (2014)
5. Gabrielli, P., Gazzani, M., Martelli, E., et al.: Optimal design of multi-energy systems with seasonal storage. Appl. Energy **219**, 408–424 (2018)
6. Clegg, S., Mancarella, P.: Integrated electrical and gas network flexibility assessment in low-carbon multi-energy systems. IEEE Transactions on Sustainable Energy **7**(2), 718–731 (2015)
7. Fabrizio, E., Corrado, V., Filippi, M.: A model to design and optimize multi-energy systems in buildings at the design concept stage. Renew. Energy **35**(3), 644–655 (2010)
8. Wen, Y., Qu, X., Li, W., et al.: Synergistic operation of electricity and natural gas networks via ADMM. IEEE Trans. Smart Grid **9**(5), 4555–4565 (2017)
9. Martinez-Mares, A., Fuerte-Esquivel, C.R.: A unified gas and power flow analysis in natural gas and electricity coupled networks. IEEE Trans. Power Syst. **27**(4), 2156–2166 (2012)
10. Qiao, Z., Guo, Q., Sun, H., et al.: An interval gas flow analysis in natural gas and electricity coupled networks considering the uncertainty of wind power. Appl. Energy **201**, 343–353 (2017)
11. Zhang, X., Shahidehpour, M., Alabdulwahab, A., et al.: Hourly electricity demand response in the stochastic day-ahead scheduling of coordinated electricity and natural gas networks. IEEE Trans. Power Syst. **31**(1), 592–601 (2015)
12. Chen, Y., Wei, W., Liu, F., et al.: A multi-lateral trading model for coupled gas-heat-power energy networks. Appl. Energy **200**, 180–191 (2017)
13. Chen, Y., Wei, W., Liu, F., et al.: Energy trading and market equilibrium in integrated heat-power distribution systems. IEEE Trans. Smart Grid **10**(4), 4080–4094 (2018)
14. Liu, X., Wu, J., Jenkins, N., et al.: Combined analysis of electricity and heat networks. Appl. Energy **162**, 1238–1250 (2016)
15. Dai, Y., Chen, L., Min, Y., et al.: Dispatch model of combined heat and power plant considering heat transfer process. IEEE Trans. Sustain. Energy **8**(3), 1225–1236 (2017)
16. Zamzam, A.S., Dall'Anese, E., Zhao, C., et al.: Optimal water–power flow-problem: Formulation and distributed optimal solution. IEEE Trans. Control Netw. Syst. **6**(1), 37–47 (2018)
17. Fooladivanda, D., Taylor, J.: Energy-optimal pump scheduling and water flow. IEEE Trans. Control Netw. Syst. **5**(3), 1016–1026 (2017)
18. Gan, L., Topcu, U., Low, S.: Optimal decentralized protocol for electric vehicle charging. IEEE Trans. Power Syst. **28**(2), 940–951 (2012)
19. Liu, Z., Wen, F., Ledwich, G.: Optimal planning of electric-vehicle charging stations in distribution systems. IEEE Trans. Power Deliv. **28**(1), 102–110 (2012)
20. Tushar, W., Saad, W., Poor, H., et al.: Economics of electric vehicle charging: A game theoretic approach. IEEE Trans. Smart Grid **3**(4), 1767–1778 (2012)
21. Liu, J., Zhang, J., Yang, Z., et al.: Materials science and materials chemistry for large scale electrochemical energy storage: from transportation to electrical grid. Adv. Func. Mater. **23**(8), 929–946 (2013)
22. Yao, S., Wang, P., Liu, X., et al.: Rolling optimization of mobile energy storage fleets for resilient service restoration. IEEE Trans. Smart Grid **11**(2), 1030–1043 (2019)
23. Rees, N.V., Compton, R.G.: Carbon-free energy: a review of ammonia-and hydrazine-based electrochemical fuel cells. Energy Environ. Sci. **4**(4), 1255–1260 (2011)
24. Li, Z., Xu, Y., Fang, S., et al.: Multiobjective coordinated energy dispatch and voyage scheduling for a multienergy ship microgrid. IEEE Trans. Ind. Appl. **56**(2), 989–999 (2019)

25. Kanellos, F.D.: Optimal power management with GHG emissions limitation in all-electric ship power systems comprising energy storage systems. IEEE Trans. Power Syst. **29**(1), 330–339 (2013)
26. Fang, S., Fang, Y., et al.: Optimal heterogeneous energy storage management for multi-energy cruise ships. IEEE Syst. J. (2020). (In press)
27. Norwegian Joy. https://www.ncl.com/in/en/cruise-ship/joy
28. Fang, S., Zhao, T., Xu, Y., et al.: Coordinated chance-constrained optimization of multi-energy microgrid system for balancing operation efficiency and quality-of-service. J. Mod. Power Syst. Clean Energy (2020). (In press)

Chapter 8
Multi-source Energy Management of Maritime Grids

8.1 Multiples Sources in Maritime Grids

8.1.1 Main Grid

The main grid plays as the main power source of land-based maritime grids since the very beginning, such as the seaports and some coastal industries. This type of maritime grids usually operates in a harbor territory and can receive electricity from the harbor city. Some equipment in those maritime grids is driven by electricity and the others may be driven by fossil fuel. Table 8.1 shows the power sources of a terminal port.

From Table 8.1, electricity, diesel, petrol and natural gas are four main power sources for a terminal port, especially the electricity and diesel, serving for most of the port-side equipment. When a seaport is less-electrified, the portion from diesel is generally higher. In recent years, the extensive electrification of seaport becomes an irreversible trend, then the electricity now has become the primary power source of a seaport. Diesel now serves for some flexible operating equipment, such as trucks and other carriers. Similar phenomena also happen in other land-based maritime grids, such as coastal factories, since when fully electrified, electricity will serve as the main energy carrier and the main grid will be the main power source.

8.1.2 Main Engines

Most types of maritime grids cannot always receive power from the main grid. They mostly operate as islanded microgrids, such as island microgrids, shipboard microgrids, and various working platforms. For the island microgrids, if they cover a wide area, a small-scale or even medium-scale power plant is possible to construct,

© The Author(s) 2021
S. Fang and H. Wang, *Optimization-Based Energy Management for Multi-energy Maritime Grids*, Springer Series on Naval Architecture, Marine Engineering, Shipbuilding and Shipping 11, https://doi.org/10.1007/978-981-33-6734-0_8

Table 8.1 Possible power sources for different equipment in a seaport (data from [1])

	Diesel	Petrol	Natural gas	Electricity
Ship-to shore cranes	•			•
Mobile cranes	•			•
Rail-mounted gantry	•			•
Rubber-tired gantry	•			•
Reach stackers	•			•
Straddle carriers	•			•
Lorries	•		•	•
Generators			•	
Building				•
Lighting				•
Reefer				•
Other vehicles	•	•	•	•

and this scenario is similar to the first case since the power plant can provide sufficient power support. For the other smaller cases, the main engines act as the main power sources instead, especially in the shipboard microgrids.

Generally, the main engines have four stages of development. The first stage is in 1900–1940, which is the initial stage of main engines. In 1910, the first diesel engine driven ship "Romagna" was launched. It uses two diesel engines manufactured by "Sulzer" company. Then in 1912, the first ocean cargo ship "Selandia" uses two DM8150x diesel engines manufactured by "B&M". In this stage, the main engines have the steamed ones and diesel ones. Then in 1940–1970, the development of main engines steps into the second stage, and this is the golden age of low-speed diesel engines. The power of a single air cylinder grows from 900–1030 kW in 1956 to 3400 kW in 1977. Then 1970–1990 is the third development stage of main engines. The theme of this stage is to reduce the fuel consumption rate. In this stage, the unit fuel consumption has reduced to 0.155–0.160 kg/(kWh), and the energy efficiency can be up to 55%. Then after 2000, the fourth stage, main engines become smarter and various advanced monitoring equipment is integrated to achieve automatic control.

Nowadays, main engines have different scales, from kilowatt to megawatt, which uses diesel, natural gas, ammonia, and so on. Some of them can use more than two types of fuels, referred to as "multi-fuel engines". Currently, main engines serve as the main power sources for many maritime grids.

8.1.3   Battery and Fuel Cell

In Chaps. 1 and 5, the energy storage technologies into maritime grids, especially the battery, are illustrated in detail. Battery stores energy in the electrochemical form and the battery cells are connected in series or in parallel or both to make up the desired voltage and capacity. There are currently many cases of battery integrated ships. Some of them are shown in Table 8.2. Nowadays, battery mostly serves as auxiliary equipment to shave the peak load of ships and benefit the operation of shipboard microgrid. In the future, the battery integration into maritime grids will be more convenient and the large-scale integration will be reality.

Since no combustion process, fuel cell has higher power generation efficiency than the traditional internal combustion engine, which is a promising power source technology in the future. Table 8.3 shows some cases of fuel cell integrated ships.

Both of battery and fuel cell have no combustion process, and are highly efficient, which are promising for future usages.

Table 8.2 Cases of battery into ships

Name	Ship types	Battery capacity	References
Ampere	Ferry	1040 kWh	[2]
Norled	Ferry	1400 kWh	[3]
Puffer	Cargo ship	2400 kWh	[4]
Princess Benedikte	Cruise ship	2.6 MWh	[5]
Elektra	Hybrid ferry	1040 kWh	[6]
Tycho Brahe	Hybrid ferry	460 kWh	[7]
Deep ocean 01	OSV	2.8 MWh	[8]
Selbjørnsfjord	Cruise ship	585 kWh	[9]
Schleswig-Holstein	Cruise ship	1.6 MWh	[10]

Table 8.3 Projects of some selected fuel cell-based ships

Ship	Power	Fuel	References
Viking Lady	330 kW	LNG	[11]
SF-Breeze	100 kW	Hydrogen	[12]
PA-X-ELL	30 kW	Methanol	[13]
MV Undine	250 kW	Methanol	[14]
US SSFC	2.5 MW	Diesel	[15]
MC-WAP	500 kW	Diesel	[16]
MS Forester	100 kW	Diesel	[17]
212 submarine U31	330 kW	Hydrogen/Methanol	[18]
212 submarine U32	240 kW	Hydrogen/Methanol	[19]
S-80 Submarine	300 kW	Ethanol	[20]

8.1.4   Renewable Energy and Demand Response

In Chaps. 1 and 5, renewable energy integration into maritime grids has been illustrated. The following Fig. 8.1 shows renewable energy integration into a seaport. Wind power, solar energy, and the main grid supply the energy demand of seaport. The ships can charge or use cold-ironing power when berthed in a seaport, which can be also viewed as using renewable energy for propulsion.

However, renewable energy is highly fluctuating and less controllable. In conventional operation patterns, the generation-side should follow the trend of renewable energy or renewable energy has to be curtailed [21]. To mitigate this issue, the demand-side can be adjusted to follow the trend of renewable energy, then the operating burden of the generation-side can be greatly reduced and the total system benefits can be improved.

In literature, demand-side management has been used in power system operation [22, 23], unit commitment [24], and so on. In the energy market, the demand-side management sources can be aggregated as one unit and acting as a "virtual power plant". In maritime grids, demand-side management is used to adjust the propulsion system of AES [25]. Later in [26], demand-side management is used to mitigate the fluctuations of photovoltaic energy. Then [27] proposes a robust demand-side management method for a photovoltaic integrated AES.

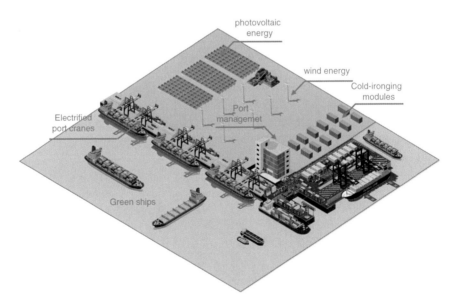

Fig. 8.1 Renewable energy integration into a seaport

8.2 Coordination Between Multiple Sources in Maritime Grids

From above, maritime grids involve multiple sources, including both generation-side and demand-side, and different sources should be coordinated to achieve better system behaviors. The coordination framework is shown as the following Fig. 8.2.

From Fig. 8.2, maritime grids consist of 4 main parts, (1) generation-side, including the main grid, main engines, fuel cell, various renewables, and so on; (2) demand-side, including the propulsion in ships, and port cranes and vehicles in a seaport, and all the load demand in different platforms; (3) Energy storage, including battery, flywheel and all the energy storage technologies can be used in maritime grids, and it should be noted that energy storage can change its roles between generation-side and demand-side, i.e., it is generation-side when discharging and it is demand-side when charging; (4) Multiple networks, including electrical, heat/cooling, water, and transportation networks, and those networks are used to deliver multiple energy flows from the generation-side to the demand-side. The energy storage and networks are the interfaces between generation-side and demand-side, thus the operating strategies of them can improve the flexibility of maritime grids.

In summary, maritime grids are a series of microgrids that have specific maritime load demand, and their operation strategies can be derived from the conventional land-based microgrids while addressing some specialties.

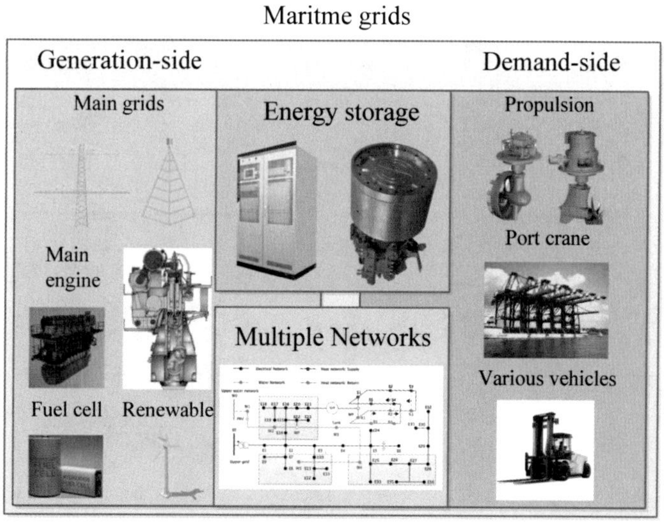

Fig. 8.2 Coordination framework of maritime grids

8.3 Some Representative Coordination Cases

8.3.1 Main Engine—Battery Coordination in AES

A single line diagram of AES is shown in Fig. 8.3 with two buses. 4 DGs are integrated into two buses. In this AES, bus A and bus B are both DC, and the DGs are all AC generators. The load demands include electric propellers and AC loads. Batteries are installed in two buses to act as auxiliary equipment.

Three sources are participating in the operation of AES, i.e., DGs, batteries, and the propulsion system of AES. The reason for the propulsion system to participate in demand response is shown as Fig. 8.4a, b.

In Fig. 8.4a, the propulsion load is cubically increasing with the cruising speed until the "wave wall". In Fig. 8.4b, the constant speed and variable speed both sail 30 nm in 6 h, but they have different load curves. In this sense, the propulsion system can adjust its load demand to coordinate with the DGs and battery to facilitate the operation of AES [25] has studied this topic and the main results are shown in Fig. 8.5a, b.

From the above Fig. 8.5a, b, the coordinated adjustment of propulsion and ESS can make the operating cost and EEOI smoother since it can mitigate the peak-valley difference of onboard power demand, which proves the effects of multi-source management on AES.

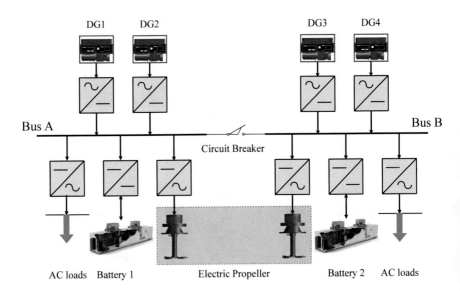

Fig. 8.3 Single-line diagram of an AES

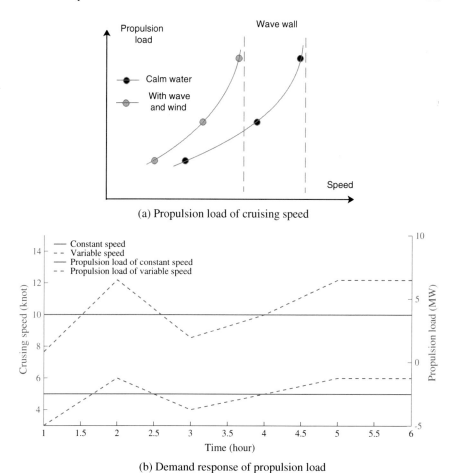

(a) Propulsion load of cruising speed

(b) Demand response of propulsion load

Fig. 8.4   Reason for propulsion system in demand response

8.3.2   Main Engine-Fuel Cell Coordination in AES

Compared with the main engines, fuel cell has smaller capacity and scale, which is suitable to undertake some small-scale load demands. Compared with the battery, fuel cell doesn't need charging, which can undertake long-term load demand [28] has studied this topic and compared two cases: (1) main engine; and (2) main engine-fuel cell. The testbed used in this study consists of a hybrid power source with the combined capacity of 180 kW (100 kW fuel cell, 30 kW battery, and 50 kW diesel generator). The results are shown in Fig. 8.6a, b. From the above curves, the integration of fuel cells can greatly reduce fuel consumption and CO_2 emission.

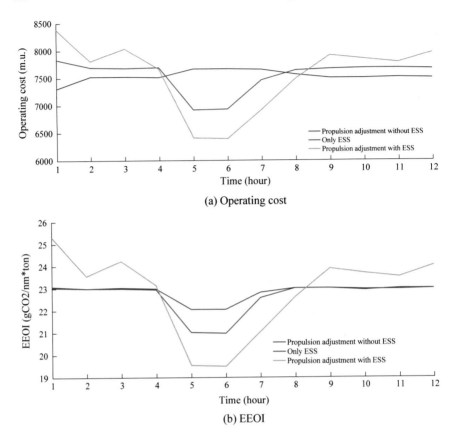

(a) Operating cost

(b) EEOI

Fig. 8.5 Operating cost and EEOI with/without ESS

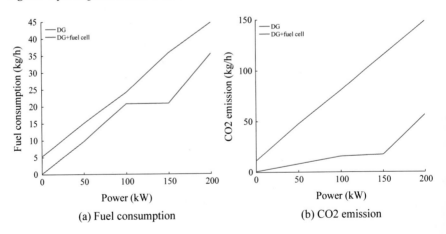

(a) Fuel consumption   (b) CO2 emission

Fig. 8.6 Comparisons between DG and DG + fuel cell

8.3.3 Demand Response Coordination Within Seaports

Chapter 6 has illustrated the operation steps of quay crane (QC). Original Fig. 6.10 is now re-drawn as Fig. 8.7 below. A typical working process of a port crane includes (1) hoist, or beginning to lift up; (2) lifting up speedily; (3) lifting up speedily and the trolley moving forward; (4) lifting up with the full speed and the trolley moving forward; (5) lifting up with slowing speed and the trolley moving with full speed; (6) the trolley moving with slowing speed; (7) lifting down speedily and the trolley moving with slowing speed; (8) settling down. Step (2) and (3) usually have the biggest power demand whereas steps (6), (7) and (8) have smaller power demands.

Chapter 6 shows the integration of ESS can recover energy when lifting down the cargo. This Chapter proposes the demand response model of port crane. The dimension of QC, cargo speed, and QC power are shown in the sub-figures in Fig. 8.8. Based on Fig. 8.8, the entire lifting cargo distance is calculated as (8.1).

$$L = h_3 + (d_1 + d_2)/2 + (h_1 + h_2)/2 \qquad (8.1)$$

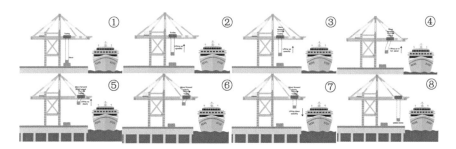

Fig. 8.7 Typical working steps for a port crane

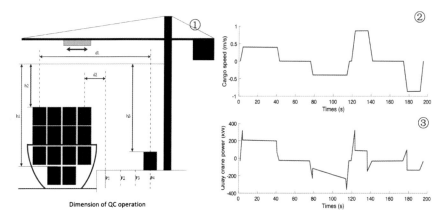

Fig. 8.8 Operation process of quay crane (QC)

From sub-figure ① and ②, the cargo speed and consumed power has a nearly linear relationship and can be shown as (8.2).

$$P = k \cdot v \qquad (8.2)$$

where P is the power of QC; k is the coefficient; and v is the cargo speed. Then the average consumed power can be shown as (8.3).

$$P_{av} \cdot T_i = \int_0^{T_i} P \, dt = k \int_0^{T_i} v \, dt = k \cdot L \qquad (8.3)$$

where P_{av} is the average consumed power of QC; T_i is the average handling time for one container. Then to handle n cargos, the consumed time is shown as (8.4).

$$T = n \cdot T_i = kL \sum_{i=1}^{n} (P_{av})^{-1} \qquad (8.4)$$

Generally there exist an upper and a lower limit on the total handling time, i.e., $T_{min} \le T \le T_{max}$. Then the demand response model of QC can be obtained.

$$\frac{T_{min}}{kL} \le \sum_{1}^{n} (P_{av})^{-1} \le \frac{T_{max}}{kL} \qquad (8.5)$$

Within the range in (8.5), the consumed power of QCs can be adjusted to facilitate the operation of seaport microgrids.

References

1. Wilmsmeier, G., Spengler, T.: Energy consumption and container terminal efficiency. Natural Resources and Infrastructure Division, UNECLAC, 2016. https://repositorio.cepal.org/handle/11362/40928?show=full
2. MV Ampere. https://en.wikipedia.org/wiki/MV_Ampere
3. Norled. https://en.wikipedia.org/wiki/Norled
4. Clyde puffer. https://en.wikipedia.org/wiki/Clyde_puffer
5. Princess Benedikte. https://www.cruisetimetables.com/cruises-from-copenhagen-denmark.html
6. The Elektra: Finland's first hybrid-electric ferry. https://ship.nridigital.com/ship_apr18/the_elektra_finland_s_first_hybrid-electric_ferry
7. MF Tycho Brahe. https://en.wikipedia.org/wiki/MF_Tycho_Brahe
8. Deep Ocean 01. http://www.sz.gov.cn/cn/xxgk/zfxxgj/tpxw/content/post_8012934.html
9. Selbjørnsfjord—Uavpic. https://uavpic.com/selbjornsfjord/
10. SMS Schleswig-Holstein. https://en.wikipedia.org/wiki/SMS_Schleswig-Holstein
11. Viking Lady. Available online: http://maritimeinteriorpoland.com/references/viking-lady/. Accessed 27 August 2018

12. SF-BREEZE. Available online: https://energy.sandia.gov/transportation-energy/hydrogen/mar kettransformation/maritime-fuel-cells/sf-breeze/. Accessed 27 August 2018
13. Pa-x-ell. Available online: http://www.e4ships.de/aims-35.html. Accessed 27 August 2018
14. METHAPU Prototypes Methanol SOFC for Ships. Fuel Cells Bull. 2008, 5, 4–5. 2859(08)70190-1
15. SFC Fuel Cells for US Army, Major Order from German Military. Fuel Cells Bull. 2012, 6, 4
16. Jafarzadeh, S., Schjølberg, I.: Emission reduction in shipping using hydrogen and fuel cells [C]// in Proceedings of the ASME International Conference on Ocean, Offshore and Arctic Engineering, Trondheim, Norway, 25–30 June 2017; p. V010T09A011
17. MS Forester. Available online: https://shipandbunker.com/news/emea/914341-fuel-cell-techno logysuccessfully-tested-on-two-vessels. Accessed 27 August 2018
18. A Class Submarine. Available online: http://www.seaforces.org/marint/German-Navy/Submar ine/Type-212A-class.htm. Accessed 27 August 2018
19. SSK S-80 Class Submarine. Available online: https://www.naval-technology.com/projects/ssk-s-80-classsubmarine/. Accessed 27 August 2018
20. Kumm, W.H., Lisie, H.L.: Feasibility study of repowering the USCGC VINDICATOR (WMEC-3) with modular diesel fueled direct fuel cells. Arctic Energies Ltd Severna Park MD: Groton, MA, USA (1997)
21. Fan, X., Wang, W., Shi, R., et al.: Analysis and countermeasures of wind power curtailment in China. Renew. Sustain. Energy Rev. 52, 1429–1436 (2015)
22. Medina, J., Muller, N., Roytelman, I.: Demand response and distribution grid operations: opportunities and challenges. IEEE Transactions on Smart Grid 1(2), 193–198 (2010)
23. Wang, Y., Pordanjani, I.R., Xu, W.: An event-driven demand response scheme for power system security enhancement. IEEE Trans. Smart Grid 2(1), 23–29 (2011)
24. Zhao, C., Wang, J., Watson, J.P., et al.: Multi-stage robust unit commitment considering wind and demand response uncertainties. IEEE Trans. Power Syst. 28(3), 2708–2717 (2013)
25. Kanellos, F.D., Tsekouras, G.J., Hatziargyriou, N.D.: Optimal demand-side management and power generation scheduling in an all-electric ship. IEEE Trans. Sustain. Energy 5(4), 1166–1175 (2014)
26. Lan, H., Wen, S., Hong, Y.Y., et al.: Optimal sizing of hybrid PV/diesel/battery in ship power system. Appl. Energy 158, 26–34 (2015)
27. Fang, S., Xu, Y., Wen, S., et al.: Data-driven robust coordination of generation and demand-side in photovoltaic integrated all-electric ship microgrids. IEEE Trans. Power Syst. 35(3), 1783–1795 (2019)
28. Roh, G., Kim, H., Jeon, H., et al.: Fuel consumption and CO_2 emission reductions of ships powered by a fuel-cell-based hybrid power source. J. Marine Sci. Eng. 7(7), 230 (2019)

Chapter 9
The Ways Ahead

9.1 Future Maritime Grids

To illustrate the future maritime grids, we re-draw Fig. 4.1 here and give a more detailed illustration for future maritime grids. The following Fig. 9.1 is renamed as "future maritime grids".

In Fig. 9.1, the main types of maritime grids including harbor city grid (2), seaport microgrids (4), offshore platforms (10), shipboard microgrids (12), offshore wind farms (14), island microgrid (15).

In the first place, harbor city grid (2) is the core and acts as the main grid for the rest of maritime grids. The main functions include receiving the land-based renewable generation (1), supplying the industrial facilities (9), providing power to seaport microgrids (4), and operating two-way ferries (12) to island microgrid (15). The former four are energy connections and the fifth is a transportation connection.

Then seaport microgrid (4) is the network within a seaport, and this microgrid receives electricity from the harbor city grid (2) and providing raw materials to the industrial facilities (9). The seaport microgrid also receives energy from the seaport renewable (6). Seaport provides berth positions to the cargo ships (16), and handling the cargos by the port cranes (13). The cargos are then lifting by the transferring vehicles (5) to the stackyard (8), and the cold-chain containers are stored in the reefer area (7). Besides, seaport microgrid provides cold-ironing power to the shipboard microgrid (12).

The offshore platforms (10) include oil drilling platforms or other construction ships. They produce raw materials and transmit them to the industrial facilities (9) or island (15) by the oil pipes or other networks. The raw materials can be also transported by cargo ships (16).

The shipboard microgrid (12) is the network installed in cargo ships (16), offshore support vessels (3), and other ships. It receives the cold-ironing power from the seaport microgrid (4), and it periodically sails between seaport and islands (15) or other places to transfer cargos.

S. Fang and H. Wang, *Optimization-Based Energy Management for Multi-energy Maritime Grids*, Springer Series on Naval Architecture, Marine Engineering, Shipbuilding and Shipping 11, https://doi.org/10.1007/978-981-33-6734-0_9

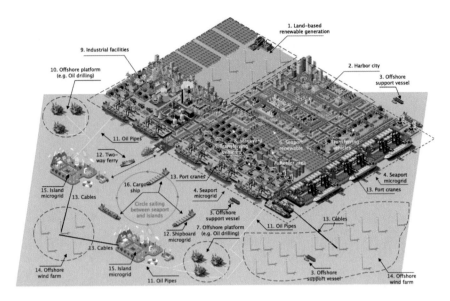

Fig. 9.1 Future maritime grids

Offshore wind farm (14) is to harvest wind energy on the sea. It has underground cables (13) to connect with the seaport (4) and then with the harbor city (2). It can also support the energy for the island microgrid (15). The offshore support vessels (3) are used to construct and repair offshore wind farms.

Island microgrid (15) is the microgrid within an island, which involves various renewable energy and other distributed generations. The scale of island microgrid depends on the area of island, and large island microgrid may have environmental agriculture facilities [1]. Island microgrid can receive the raw materials from the offshore platforms, and exchange materials with the seaport (4) by cargo ships (16). The tourists can have two-way traveling between islands and harbor city by two-way ferries (12).

From above, maritime grids undertake different maritime tasks and they are tightly coupled and they should be studied as one unit. Some typical operating scenarios are important and shown below.

(1) The coordination between the seaport microgrid and the harbor city grid. In this scenario, the harbor city grid is the main grid, and the seaport microgrid purchases electricity from the main grid to support the within equipment, i.e., port cranes, transferring vehicles, reefer area, and so on [2–4] have studied this scenario.

(2) The coordination between the seaport microgrid and the shipboard microgrids. In this scenario, the seaport allocates berth positions to the berthed-in ships and providing cold-ironing power and logistic services [5, 6] have studied this scenario.

(3) The coordination between the shipboard microgrids and the island microgrids. This scenario is similar to the case between seaport and ships when an island has a very strong power network. When the power grid of the island is weak, the ships may in reverse support the islands, which is referred to as "mobile power plant" [7].

(4) The coordination between the offshore platforms. There are generally many offshore platforms in an ocean area, and they should coordinate with each other to complete the same task, i.e., oil drilling, construction, and so on.

9.2 Data-Driven Technologies

9.2.1 Navigation Uncertainty Forecasting

Navigation uncertainty generally comes from uncertain weather, and Chap. 4 has emphasized the influences of navigation uncertainty on the operation of maritime grids. Until now, there are many data-driven maritime weather forecasting methods for ships and seaport [8–10], in different timescales, or by different algorithms, using different attributes, and also have different advantages and disadvantages. Our focus is on how to use those forecasting datasets to generate the distributions and uncertainty sets of energy management models. With the obtained distributions or uncertainty sets, stochastic and robust programming models can be formulated for different operating scenarios.

In recent research [11], a novel data-driven heuristic framework for vessel weather routing is formulated as Fig. 9.2. Based on the weather forecasting results, the ship chooses a better sailing route to save fuel consumption. The main key performance indicators (KPIs) of ships can also be predicted.

Fang et al. [12] also studies the robust energy management of all-electric ships when considering navigation uncertainties, but the weather conditions are simply classified as four sub-scenarios and only the worst case is considered. In the future, more accurate uncertainty sets should be forecasted to facilitate the operation of maritime grids.

9.2.2 States of Battery Energy Storage

Chapters 5–8 have emphasized the critical roles of battery energy storage in the maritime grids for load leveling and power quality issues. Generally, there are six states for battery energy storage, i.e., state of charge (SOC), state of power (SOP), state of energy (SOE), state of safety (SOS), State of temperature (SOT), and state of health (SOH). The above states are all essential indicators for the battery management system and many methods have been proposed to estimate them, and various data-driven techniques have been utilized.

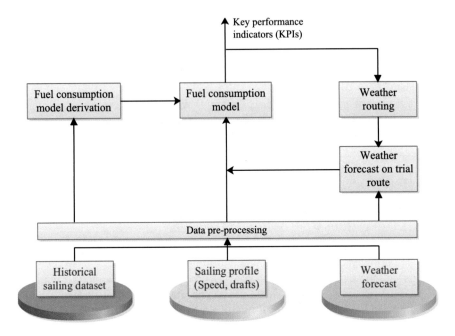

Fig. 9.2 Flowchart of the data-driven weather routing method

Generally, SOC is defined as the ratio of available capacity to the nominal capacity. Here the nominal capacity stands for the maximum amount of charge. Using the tank of a fuel vehicle as an analogy, SOC is similar to the fuel gauge. The definition of SOC is shown in (9.1) [13].

$$SOC(t) = SOC(t_0) + \int_{t_0}^{t} I(t) \cdot \eta / Q_n dt \qquad (9.1)$$

where $I(t)$ is the current of battery energy storage; Q_n is the nominal capacity; η is the coulombic efficiency.

Another indicator, SOP is generally defined as the available power that a battery can supply to or absorb over a time horizon [14]. The definition of SOP is shown as (9.2).

$$\begin{cases} SOP^{charge}(t) = max\left(P_{min}, V(t+\Delta t) \cdot I_{min}^{charge}\right) \\ SOP^{discharge}(t) = min\left(P_{max}, V(t+\Delta t) \cdot I_{max}^{discharge}\right) \end{cases} \qquad (9.2)$$

where P_{min} and P_{max} are the lower and upper limits of power; I_{min}^{charge} and $I_{max}^{discharge}$ are the lower and upper limits of current.

Another indicator, SOE is defined as the supplying/absorbing discrepant energy amounts in different voltage levels, which is given as (9.3).

$$SOE(t) = SOE(t_0) + \int_{t_0}^{t} P(t) / E_N dt \tag{9.3}$$

where $P(t)$ is the power; E_N is the nominal energy capacity.

Another indicator, SOS represents the hazard level when battery operating, and the definition is given as (9.4).

$$H_r = H_s \cdot H_l \tag{9.4}$$

where H_r, H_s, H_l represent the hazard risk, hazard severity, and hazard likelihood, respectively. In [15], H_s can vary from 0 to 7 as an integer to represent the hazard level; H_l can take values from 1 to 10 to represent the occurrence percentage of failures; H_r utilizes two states (i.e., H_s and H_l) to find a safe operating region.

The temperature has been recognized as one main factor for battery degradation, and the SOT indicates the operating temperature of battery, including the estimations of external, internal, and temperature distribution. In general, the external temperature is easy to control, and the internal temperature and temperature distribution are much more important to represent the state of battery. The estimation of SOT is based on the thermal dynamic model as (9.5) [16].

$$\begin{cases} C_C \cdot \dot{T}_c = \dot{Q} + (T_s - T_c) / R_x \\ C_S \cdot \dot{T}_S = (T_\infty - T_s) / R_u + (T_S - T_c) / R_c \end{cases} \tag{9.5}$$

where T_s and T_c are the surface and core temperature, respectively; R_u and R_c are the conductive and convective resistances, respectively; T_∞ is the ambient temperature. The last indicator is the SOH to represent the health state of battery, which is given by the following.

$$SOH = C_a / C_r \times 100\% \tag{9.6}$$

where C_a and C_r are the actual and rated capacity, respectively.

There are many estimation methods for the above six states of battery energy storage [13–20], and these methods belong to multiple timescales, which are shown as Fig. 9.3 below.

Besides, there are different timescales for each state. For example, there are offline training and online estimation stages for SOH estimation in Fig. 9.4 [21].

In summary, current state estimation methods can be used in maritime grids when addressing the working conditions of highly humid and saline, and high-temperature. In Fig. 9.4, these characteristics should be considered in the experimental conditions and the uncertainty management of SOH estimation model. However, there is still very little literature on this topic now.

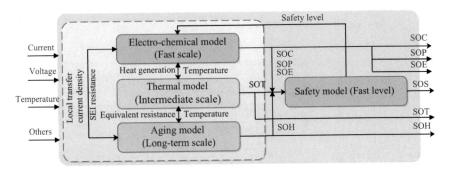

Fig. 9.3 Time-scale of state estimation of battery

Besides, the above state estimation methods are for a single battery cell. As shown in Fig. 9.5, a battery pack is comprised of many battery cells and generally different cells have different degradation speeds. This difference should be considered, named as the inconsistency of state estimation.

9.2.3 Fuel Cell Degradation

The importance of fuel cells in maritime grids has been clarified in Chap. 8, and the technological development will drive the further large-scale integration of fuel cells. Similar to battery, the degradation of fuel cells is important and certain methods should be proposed to estimate the degradation in different scenarios. Generally, the fuel cell degradation methods can be classified as (1) stack voltage degradation model; (2) Electrochemical impedance spectrometry (EIS) impedance estimation; (3) Remaining useful life (RUL) estimation. Their advantage and disadvantages are shown in Table 9.1.

The stack voltage degradation models use the output voltage V_{stack} to demonstrate the degradation phenomenon, and are usually based on two prototypes, shown in (9.7) and (9.8), respectively.

$$\begin{cases} V_{stack} = V_{rate} \cdot D_{fc} \\ D_{fc} = k_p \cdot (P_1 \cdot n_1 + P_2 \cdot n_2 + P_3 \cdot t_1 + P_4 \cdot t_2) \end{cases} \tag{9.7}$$

$$V_{stack} = V_0 - b \cdot \log(i_{fc}) - r \cdot i_{fc} + \alpha \cdot i_{fc}^{\sigma}(1 - \beta \cdot i_{fc}) \tag{9.8}$$

In (9.7), V_{stack} is the stack voltage; D_{fc} is the degradation rate; k_p is the accelerating coefficient; P_1, P_2, P_3 and P_4 are the degradation rates led by the load change, start-up/shut-down, idling, and high-power demand, respectively; and n_1, n_2, t_1, t_2 denotes the corresponding times/time-periods. In (9.8), V_0 represents the open-circuit voltage; i_{fc} is the current of fuel cell; b, r, α, and σ are parameters deduced from the experiment dataset. When the dataset changes, all the parameters should be adjusted.

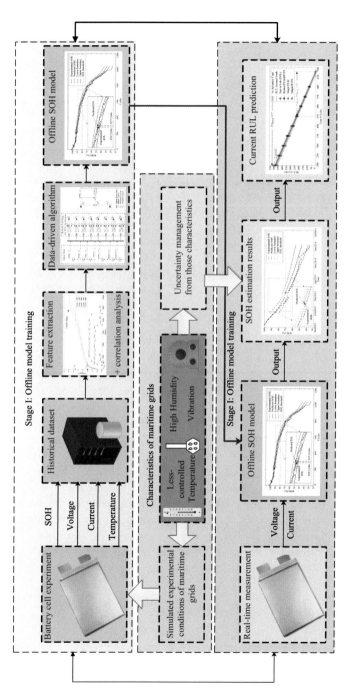

Fig. 9.4 Time-scale of state estimation of battery

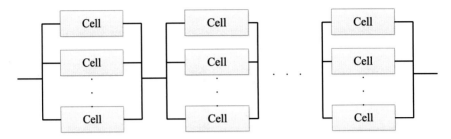

Fig. 9.5 Battery cells and Battery pack

Table 9.1 Summary of different fuel cell degradation methods

Model	References	Methods	Advantages	Disadvantages
Stack voltage degradation model	[22–24]	data-driven parameter recognization	1. Easily implement 2. Less requirement on theoretical analysis	1. Highly rely on experiment 2. Hard to adjust the parameters
EIS impedance estimation	[25, 26]	Model-based methods	1. Easily implement 2. Suitable for diagnostics field	Cannot directly forecast SOH
RUL estimation	[27–29]	Hybrid methods	Robustness to uncertainties	Computational stress

EIS is carried out by adding a small sinusoidal perturbation on the nominal current and then the EIS impedance can be calculated as the ratio between the response and the perturbation. This method can characterize the phenomenon inside the fuel cell and evaluate the fuel cell degradation [25], which are widely used in the diagnostics field, but it cannot give the information of SOH. RUL methods are a series of hybrid methods, which can be based on the semi-empirical model [28], or various machine-learning methods [30]. Since the recent development of data-mining techniques, RUL methods also have many new applications [30].

In summary, the fuel cell degradation estimation is similar to the battery and a similar estimation process can be utilized. The gaps before implementing in maritime grids are addressing the working conditions with high humidity, and high-temperature. However, there is still very little literature working on this topic.

9.2.4 Renewable Energy Forecasting

Chapter 5 has emphasized the importance of renewable energy forecasting of maritime grids. Figures 5.10 and 5.11 show that the forecasting of renewables onboard should consider more features. To recall this part, Figs. 5.10 and 5.11 are re-drawn as Fig. 9.6a, b as follows.

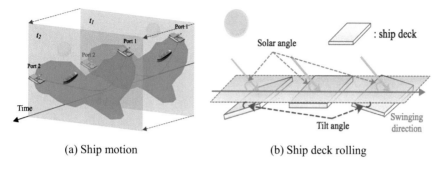

      (a) Ship motion   (b) Ship deck rolling

Fig. 9.6 Two extra features in onboard renewable energy forecasting [31]

An adaptive clustering method for onboard photovoltaic energy is proposed in [32]. The sketch process is shown in Fig. 9.7.

With the proposed method, the scenarios of photovoltaic energy can be adaptively obtained, and the administrator can give an optimal energy scheme for each scenario. Later in [33], the ship motion, temperature, irradiance, and temperature are all considered and a hybrid ensemble forecasting method is formulated as Fig. 9.8.

With the proposed method in Fig. 9.8, the onboard photovoltaic energy can be predicted with more accuracy. Two representatives above show the keys for the renewable energy forecasting in maritime grids: (1) properly clustering the original dataset, and the main reason is the weather conditions may change more frequent in maritime grids than other land-based applications; (2) putting more practical features into the forecasting model, such as the ship motion and rolling. With the development of renewable energy technology, the penetration of large-scale renewable energy into maritime grids will become reality, and the renewable energy forecasting in maritime grids will find a promising scenario for application.

9.3 Siting and Sizing Problems

9.3.1 Energy Storage Integration

Chapter 6 has clarified the functions of energy storage in the long-term operation of maritime grids: (1) improving economic and environmental characteristics of maritime grids [5, 12, 34]; (2) benefiting the operation of onboard equipment [31, 32, 35]; (3) improving the resilience of maritime grids [36], which are illustrated in Fig. 9.9.

In Fig. 9.9a, the main engines and energy storage are sharing the highly fluctuated power demand via maritime grids. The energy storage shares the highly fluctuated part and the main engines can work in a constant and economic working condition. In Fig. 9.9b, new equipment is integrated into the maritime grid, and the energy storage

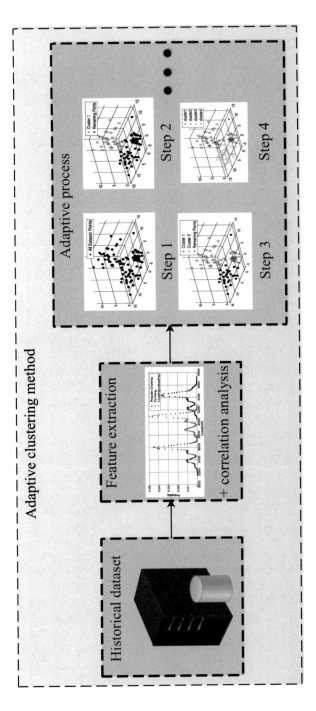

Fig. 9.7 Adaptive clustering methods for onboard photovoltaic energy

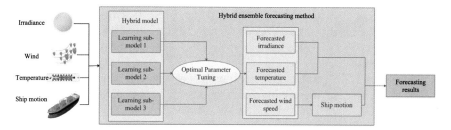

Fig. 9.8 Hybrid ensemble forecasting method

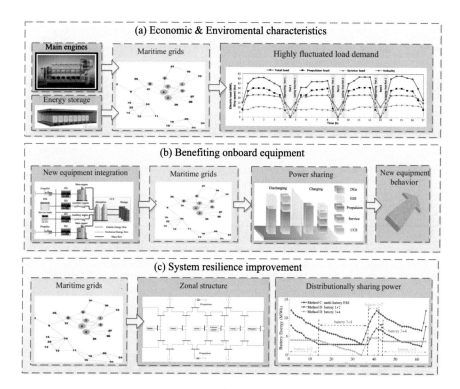

Fig. 9.9 Main functions of energy storage in maritime grids

can share the power demand of new equipment to improve its behavior. In Fig. 9.9c, energy storage is installed distributionally in different zones of maritime grid, and energy storages in different zones share the power demand, and make the system be resilient to various failures.

Since the important functions above, energy storage gradually becomes essential equipment in maritime grids to improve system characteristics. However, energy

storage, generally battery for long-term operation, is still expensive and the installment area is also another limit for energy storage. The balance between the economic
benefits and the system characteristics motivates the siting and sizing problems of
energy storage.

Reference [32, 37, 38] propose optimal energy storage sizing methods after
comprehensively studying the influences of energy storage on the penetration of
photovoltaic energy into maritime grids, which considers effects of the ship motion,
deck rolling, and solar irradiation density. In seaport, [2] proposes six indexes to indicate the green operation, and a two-stage energy storage sizing problem is formulated
to improve the indexes. Since the battery is limited in power density, [34, 39] propose
optimal sizing methods for hybrid energy storage, i.e., high power density energy
storage for the high-frequency load demand, and battery for the low-frequency load
demand. For the system resilience, a distributed energy storage siting and sizing
model is formulated, and the simulation results show that the distributionally installed
energy storages benefit the resilience.

In summary, future research should consider more specialties of maritime grids,
which are shown as follows.

(1) Special network structures. Maritime grids have a different network structure
     compared with conventional land-based microgrids. This feature in ships has
     been illustrated in Chap. 5 as Fig. 5.11. We re-draw this figure to Fig. 9.10
     below, and we can find the network structure of ships is zonal and parallelly
     designed.
(2) The distributional installment of energy storage. Different from the land-based
     applications, the energy storages in maritime grids are mostly distributionally
     installed. For example, the energy storage system in ships is usually separated
     into several parts and installed in different watertight compartments for system

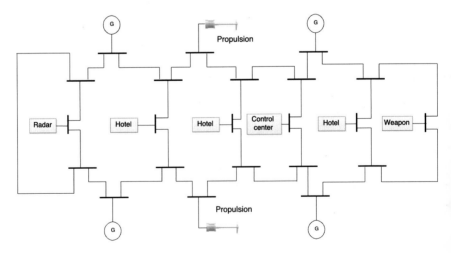

Fig. 9.10 The graph topology of an all-electric ship

resilience. In seaport, energy storages have different functions, i.e, for cold-ironing, for port cranes, for electric truck charging, and so on. So the energy storages also need to be distributionally installed.

(3) The redundant capacity of energy storage. Different from conventional land-based microgrids, maritime grids generally receive less support from the main grid. In this sense, energy storage is viewed as one of the main power sources to improve system resilience, and therefore needs to have a redundant capacity.

9.3.2 Fuel Cell Integration

Chapter 8 has revealed the fuel cell is a promising power source for the future maritime grids, and its integration is an irreversible trend. Currently, there are many practical cases and studies on the siting and sizing of fuel cells in maritime grids. With these cases, fuel cell shows similar effects as the integration of energy storage, i.e., highly flexible, energy-efficient, no combustion process, and easily maintained. The functions are also similar: (1) improving economic and environmental characteristics of maritime grids; (2) benefiting the operation of onboard equipment; (3) improving the resilience of maritime grids. Although these similarities, fuel cell is a power source and has no need to charge, and therefore the fuel cell is able to sustain the long-term power demand.

As above, future research should consider the following aspects as Fig. 9.11 before it can integrate into maritime grids.

(1) Fuel cell is a power source and has similar functions with the main engines. In this sense, the maritime grids should be expanded for its integration, i.e., structure modification.

(2) Generally, fuel cell and main engines serve different load demands, i.e., main engines for the large-scale load demand such as propulsion, and fuel cell for the small-scale but critical load demand such as control center. The division of responsibilities should be considered.

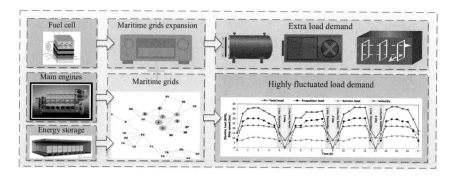

Fig. 9.11 Fuel cell integration and maritime grid expansion

9.4 Energy Management

With the above illustrations, the main target for maritime grids is to achieve the cost-efficient and green development of the maritime industry, and the energy management methods/strategies are fundamental for this target. In the future, the energy management of maritime grids should have two main abilities: (1) Ambient environment perception, i.e., the real-time perception of working conditions and the quick responding abilities for the changes of working conditions. (2) Optimal energy scheduling ability, i.e., real-time perception of system conditions and the ability for the optimal energy scheduling among different sources and equipment. These two abilities are shown in Fig. 9.12 below.

From Fig. 9.12, the first ability, ambient environment perception, relies on real-time data measurement and the corresponding data-driven techniques. This ability can provide adequate inputs to indicate the energy scheduling of maritime grids. It should be noted that the ambient environment includes the working conditions and the coordination from other maritime grids, such as the coordination between berthed-in ships and seaport.

Then the second ability, optimal energy scheduling ability, should integrate all the current management methods, i.e., the methods mentioned in Chaps. 5–8, namely, uncertainty management, energy storage management, multi-energy management, and multi-source energy management, and determines an optimal energy scheme to respond to the ambient working conditions.

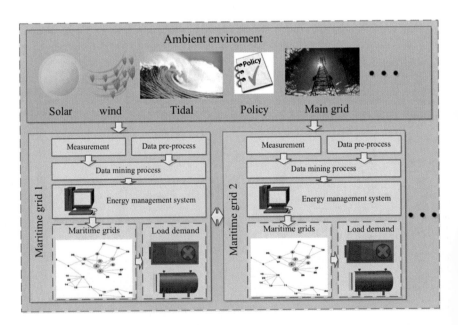

Fig. 9.12 Energy management of maritime grids

9.5 Summary

Generally, maritime grids are born under the trend of maritime transportation electrification, and this trend is irreversible in the future. From the views of electrical engineering, maritime grids are a series of microgrid-scale networks which undertake different maritime tasks. The electrical network serves as the backbone and connects with other networks with different functionalities. This characteristic determines the operation of maritime grids should have plenty of similarities with conventional microgrids, but the maritime tasks involved further make the maritime grids with many distinguishing features. In this sense, it is essential and also very necessary to study this type of special microgrids before they can be implemented in real-world.

In this book, maritime grids are defined as those networks installed in harbors, ports, ships, ferries, or vessels. A typical maritime grid consists of generation, storage, and critical loads, and can operate either in grid-connected or in islanded modes, and operate under both the constraints of the energy system and maritime transportation system, and formulates as a "maritime multi-energy system", and the energy management of this special system will shape the energy efficiency improvement of the future maritime transportation system.

In this book, optimization-based energy management methods are comprehensively reviewed and overviewed with plentiful case studies. In Chaps. 1–4, i.e., (1) the introduction for maritime grids, (2) the mathematical basics of optimization; (3) mathematical formulation of management targets and (4) formulation and solution of maritime grid optimization, give illustrative descriptions on the research focus. Then in Chaps. 5–8, four aspects, i.e., (1) energy management under uncertainties, (2) energy storage management, (3) multi-energy management, and (4) multi-source energy management, are discussed. At last, this chapter overviews the future roadmap in four parts, i.e., (1) future maritime grids, (2) data-driven technologies, (3) siting and sizing problems, and (4) energy management. With the above arrangement, the initial research framework of maritime grids has been launched and specific efforts are expected in this field for future development.

References

1. Walsh, K., O'Sullivan, D., Young, R. et al.: Medieval land use, agriculture and environmental change on Lindisfarne (Holy Island), Northumbria. Ecological relations in historical times: human impact and adaptation, pp. 101–21 (1995)
2. Molavia, A., Shib, J., et al.: Enabling smart ports through the integration of microgrids: a two-stage stochastic programming approach. Appl. Energy **258**, 114022 (2020)
3. Kanellos, F.D., Volanis, E.S., Hatziargyriou, N.D.: Power management method for large ports with multi-agent systems. IEEE Trans. Smart Grid **10**(2), 1259–1268 (2017)
4. Gennitsaris, S.G., Kanellos, F.D.: Emission-Aware and cost-effective distributed demand response system for extensively electrified large ports. IEEE Trans. Power Syst. **34**(6), 4341–4351 (2019)

5. Wen, S., Zhao, T., Tang, Y., et al.: A joint photovoltaic-dependent navigation routing and energy storage system sizing scheme for more efficient all-electric ships. IEEE Trans. Transp. Electrif. **6**(3), 1279–1289 (2020)
6. Fang, S., Wang, Y., Gou, B., et al.: Toward future green maritime transportation: an overview of seaport microgrids and all-electric ships. IEEE Trans. Veh. Technol. **69**(1), 207–219 (2019)
7. Tanaka, E.: Power generation plant ship: U.S. Patent Application 09/865,495. 2002–12-5
8. Georgiana, E.: A decision tree for weather prediction. Buletinul. LXI **1**, 77–82 (2009)
9. Sawale, G.J., Gupta, S.R.: Use of artificial neural network in data mining for weather forecasting. Int. J. Comput. Sci. Appl. **6**(2), 383–387 (2013)
10. Shanmuganathan, S., Sallis, P.: Data mining methods to generate severe wind gust models. Atmosphere **5**(1), 60–80 (2014)
11. Gkerekos, C., Lazakis, I.: A novel, data-driven heuristic framework for vessel weather routing. Ocean Eng. **197**, 106887 (2020)
12. Fang, S., Xu, Y., Wang, H. et al.: Robust operation of shipboard microgrids with multiple-battery energy storage system under navigation uncertainties. IEEE Trans. Vehic. Technol. (2020). In press
13. Hannan, M.A., Lipu, M.S., Hussain, A., et al.: A review of lithium-ion battery state of charge estimation and management system in electric vehicle applications: challenges and recommendations. Renew. Sustain. Energy Rev. **78**, 834–854 (2017)
14. Plett, G.L.: High-performance battery-pack power estimation using a dynamic cell model. IEEE Trans. Veh. Technol. **53**(5), 1586–1593 (2004)
15. Ashtiani, C.: Analysis of battery safety and hazards' risk mitigation. ECS Trans. **11**(19), 1 (2008)
16. Lin, X., Stefanopoulou, A.G., Perez, H.E. et al.: Quadruple adaptive observer of the core temperature in cylindrical Li-ion batteries and their health monitoring. In: 2012 American Control Conference (ACC), pp. 578–583. IEEE (2012)
17. Xie, J., Ma, J., Chen, J.: Available power prediction limited by multiple constraints for LiFePO4 batteries based on central difference Kalman filter. Int. J. Energy Res. **42**(15), 4730–4745 (2018)
18. Cannarella, J., Arnold, C.B.: State of health and charge measurements in lithium-ion batteries using mechanical stress. J. Power Sources **269**, 7–14 (2014)
19. Berecibar, M., Gandiaga, I., Villarreal, I., et al.: Critical review of state of health estimation methods of Li-ion batteries for real applications. Renew. Sustain. Energy Rev. **56**, 572–587 (2016)
20. Zheng, L., Zhang, L., Zhu, J., et al.: Co-estimation of state-of-charge, capacity and resistance for lithium-ion batteries based on a high-fidelity electrochemical model. Appl. Energy **180**, 424–434 (2016)
21. Gou, B., Xu, Y., Feng, X.X.: State-of-Health estimation and remaining-useful-life prediction for lithium-ion battery using a hybrid data-driven method. IEEE Trans. Vehic. Technol. (2020)
22. Pei, P., Chang, Q., Tang, T.: A quick evaluating method for automotive fuel cell lifetime. Int. J. Hydrogen Energy **33**(14), 3829–3836 (2008)
23. Chen, H., Pei, P., Song, M.: Lifetime prediction and the economic lifetime of proton exchange membrane fuel cells. Appl. Energy **142**, 154–163 (2015)
24. Xu, L., Li, J., Ouyang, M., et al.: Multi-mode control strategy for fuel cell electric vehicles regarding fuel economy and durability. Int. J. Hydrogen Energy **39**(5), 2374–2389 (2014)
25. Petrone, R., Zheng, Z., Hissel, D., et al.: A review on model-based diagnosis methodologies for PEMFCs. Int. J. Hydrogen Energy **38**(17), 7077–7091 (2013)
26. Cadet, C., Jemeï, S., Druart, F., et al.: Diagnostic tools for PEMFCs: from conception to implementation. Int. J. Hydrogen Energy **39**(20), 10613–10626 (2014)
27. Bressel, M., Hilairet, M., Hissel, D., et al.: Extended Kalman filter for prognostic of proton exchange membrane fuel cell. Appl. Energy **164**, 220–227 (2016)
28. Silva, R.E., Gouriveau, R., Jemei, S., et al.: Proton exchange membrane fuel cell degradation prediction based on adaptive neuro-fuzzy inference systems. Int. J. Hydrogen Energy **39**(21), 11128–11144 (2014)

29. Chandrasekaran, R., Bi, W., Fuller, T.F.: Robust design of battery/fuel cell hybrid systems— methodology for surrogate models of Pt stability and mitigation through system controls. J. Power Sources **182**(2), 546–557 (2008)
30. Hu, X., Feng, F., Liu, K., et al.: State estimation for advanced battery management: Key challenges and future trends. Renew. Sustain. Energy Rev. **114**, 109334 (2019)
31. Fang, S., Xu, Y., Wen, S., et al.: Data-driven robust coordination of generation and demand-Side in photovoltaic integrated all-electric ship microgrids. IEEE Trans. Power Syst. **35**(3), 1783–1795 (2019)
32. Yao, C., Chen, M., Hong, Y.Y.: Novel adaptive multi-clustering algorithm-based optimal ESS sizing in ship power system considering uncertainty. IEEE Trans. Power Syst. **33**(1), 307–316 (2017)
33. Wen, S., Zhang, C., Lan, H. et al.: A hybrid ensemble model for interval prediction of solar power output in ship onboard power systems. IEEE Trans. Sustain. Energy (2019)
34. Fang, S., Xu, Y., Li, Z., et al.: Two-step multi-objective management of hybrid energy storage system in all-electric ship microgrids. IEEE Trans. Veh. Technol. **68**(4), 3361–3373 (2019)
35. Fang, S., Xu, Y., Li, Z., et al.: Optimal sizing of shipboard carbon capture system for maritime greenhouse emission control. IEEE Trans. Ind. Appl. **55**(6), 5543–5553 (2019)
36. Lai, K., Illindala, S.: Graph theory based shipboard power system expansion strategy for enhanced resilience. IEEE Trans. Ind. Appl. **54**(6), 5691–5699 (2018)
37. Lan, H., Wen, S., Hong, Y.Y., et al.: Optimal sizing of hybrid PV/diesel/battery in ship power system. Appl. Energy **158**, 26–34 (2015)
38. Wen, S., Lan, H., Yu, D.C., et al.: Optimal sizing of hybrid energy storage sub-systems in PV/diesel ship power system using frequency analysis. Energy **140**, 198–208 (2017)
39. Hou, J., Sun, J., Hofmann, H.: Control development and performance evaluation for battery/flywheel hybrid energy storage solutions to mitigate load fluctuations in all-electric ship propulsion systems. Appl. Energy **212**, 919–930 (2018)

Printed in the United States
by Baker & Taylor Publisher Services